Crisis Management

Crisis Management
Planning and Media Relations for the Design and Construction Industry

Janine L. Reid

JOHN WILEY & SONS, INC.

New York / Chichester / Weinheim / Brisbane / Singapore / Toronto

This book is printed on acid-free paper. ∞

Published by John Wiley & Sons, Inc.
Published simultaneously in Canada.

This publication is designed to provide accurate and authoritative information in regard to the subject matter covered. It is sold with the understanding that the publisher is not engaged in rendering professional services. If professional advice or other expert assistance is required, the services of a competent professional person should be sought.

Library of Congress Cataloging-in-Publication Data
Crisis management: planning and media relations for the design and construction industry/
    Janine L. Reid
        p.   cm.
    Includes bibliographical references and index.
    ISBN 0-471-35419-8 (cloth)
        1. Construction industry—Public relations.   2. Engineering firms—Public relations.
    3. Construction industry—Accidents—Press coverage.   4. Building failures—Press coverage.
    5. Crisis management—Planning.   I. Janine L. Reid.
    HD9715.A2 C75 2000
    658.4′056—dc21   99-088401

Printed in the United States of America

10 9 8 7 6 5 4 3 2 1

For Justin

CONTENTS

FOREWORD

Planning and management are the bricks and mortar of construction. One without the other is a recipe for problems. The best of managers, working without a plan, are headed for failure, just as certainly as a plan lacking management. In real-world construction, though, the finest plan backed by top-notch managers won't guarantee success. Improve the odds? Definitely. But when you throw together dozens of trades, subs, architects, and engineers, probably for the first time, on a typical one-up project, a crisis of some sort is as certain as sunrise.

How to deal with that eventual crisis is what Janine Reid's book is all about—from crisis identification through crisis management to crisis recovery. When bad things happen to good companies, the ability to deal with the crisis is what separates truly professional firms from the wannabes.

Howard B. Stussman
Editor-In-Chief, ENR

PREFACE

Is *crisis management* an oxymoron? For the unprepared, the answer is a resounding YES! I speak from experience, because I was one of the unprepared.

It was August 18, 1985, and I was in the midst of preparing a presentation for a general contractor I was working for in Denver, Colorado, when the phone rang. Annoyed at the disruption, I picked up the receiver and answered with a sharp "Hello."

"I have just received the most frightening call of my life. Some lunatic said that he has planted a bomb on the site and it is going to explode in two hours. What should I do?" shrieked Mike, a superintendent on a project my company was working on in Albuquerque, New Mexico.

This project, which broke ground in March, was a 15-story commercial office building located in downtown Albuquerque. The job site encompassed an entire city block, and we were approximately 45% complete at the time of Mike's call.

As head of marketing, it was suddenly my responsibility. Because I did not know what to do, I did what any unprepared person would do: I panicked. "Evacuate the job site and I'll be on the next plane to Albuquerque," was my sophomoric response.

On the plane to Albuquerque, I brooded that I was truly in over my head and had no clue how to manage the situation. No one else in my company knew either, so I had nowhere to turn for counsel. The words of Lily Tomlin came to mind: "We are all in this alone." I had never before, or since, felt so alone. This bomb threat was happening on my watch, and I was not ready.

Upon my arrival in Albuquerque, my heart rate seemed to enter the danger zone. As my cab pulled up to the job site, I saw police cars and fire trucks everywhere. The workers had been evacuated and a police officer with a bomb-sniffing dog was combing the property. Once inside the job-site trailer, I immediately was immersed in a series of strategy sessions.

The bomb threats continued for three days and placed an immense amount of stress on our ad hoc crisis management team. Every time the phone rang in the trailer, the team experienced everything from paralysis to anger.

The authorities advised us that there was a high incidence of bomb threats in the downtown area and that they felt these calls fell in the "threat" category. However, there were no positive assurances. We hired 24-hour security and issued badges to everyone who worked on the project. No one was allowed near the site unless cleared by security. I stayed in the trailer playing the never-ending game of *What if?* The purpose of this game, which is described in Chapter 1, is to anticipate any and all spin-off crises of the initial crisis and to determine the appropriate strategy to counter any negative effects—for example, *What if* it is a real bomb? *What if* there is an explosion and people are hurt or killed? *What if* the media arrive and start asking questions? The *what-if* game can be endless, each potential crisis more horrifying than the last.

On the fourth day, the calls stopped. Hesitantly, everyone returned to work. Continuing our strict security measures throughout the rest of the project, we completed it on time and [barely] within budget.

Crises are excellent teachers. As I reflect on my first crisis, I treasure the accelerated education I received in Albuquerque. Two of the many lessons I learned during this incident were that no one is immune to the inevitable and that crises do not discriminate. They do not care if you have the best safety program in the business. Because of the human element involved, the best safety program is no guarantee against a crisis. Furthermore, crises do not care if you are a small and specialized company. In fact, a crisis can deal a crippling blow to a small and unprepared company with limited resources. The sky can fall on the best of companies; however, a proactive company understands that stuff can happen and so will be prepared for the inevitable.

Albuquerque made me a believer in having an established protocol to guide responses to crises, so on my flight back to Denver I began writing my first crisis management plan. It served me well because I had the opportunity to test it in several subsequent situations. The plan afforded me the luxury to be proactive rather than reactive, and when the next crisis came, it allowed me to position my company in a much more favorable light with its audiences.

In August 1985, my career changed direction. I began researching this discipline called crisis management planning and in 1987 wrote my first book, *What to Do When the Sky Starts Falling.* I have had the opportunity work on dozens upon dozens of crisis situations and am continually learning about new techniques and approaches to help a company regain its position in the marketplace after a crisis and to create a support system among its employees. Total crisis management is what this book is all about.

However, there is no one-size-fits-all protocol for every crisis. Each has its own personality and challenges. With this warning in mind, read this book. Its intent is to share research, experiences, and lessons learned from working on a host of crises over the past two decades. To facilitate this process, the book takes you through the four phases of a crisis: crisis identification and prevention, crisis management, crisis communications, and crisis recovery. The material is presented in a question and answer format to allow for quick reference.

So is *crisis management* an oxymoron? For the prepared, the answer is no. My experience has shown that your chances of managing a crisis increase exponentially with the amount of preparation and planning you do beforehand. This book provides the information you need to anticipate crises. My hope is that you never need to activate your crisis management plan, but you will sleep better just knowing that it is there—just in case.

ACKNOWLEDGMENTS

First, I wish to acknowledge the industry that has been my home for the last 20-plus years. The design and construction industry employs some of the finest individuals in the world and offers great excitement and opportunity.

To Dan Murphy, Dale Daul, and Curtis Childress at The St. Paul Construction, who believe in the discipline of crisis management. My appreciation also goes to those who have dedicated their careers to the field of safety. These individuals have provided me with tremendous support throughout the years. Their dedication to reducing risks and saving lives goes beyond being laudable.

I was honored to have Cletus Keating, Ph.D., as my editor. His gentle guidance, patience, and counsel were invaluable and very much appreciated. My thanks also goes to Mary Prendergast, who was instrumental in researching various components of this book.

Finally, thank you, Hugh Rice. When I struggled with the thought of writing another book, you reminded me that this book was already written—it was merely looking for a way to get out.

Crisis Management

1

CRISIS IDENTIFICATION
AND PREVENTION

"Why didn't we see this coming?" is the lament of many business owners as they stare into the eyes of a full-blown crisis.

This chapter will help you avoid this lament. Business would be so much easier if risks could be identified and contained before they threaten a company's survival. Although some may argue that such a vision is unrealistic in the construction industry because of its high-risk nature, I defy this "reality" and issue a challenge to everyone in the industry: Identify your risks and prevent them from becoming crises. To do so, you must first know what a crisis is.

What is a crisis?

The word *crisis* has a variety of definitions. An insurance broker once told me, "A crisis is when a contractor opens the door for business every day." True as this claim may seem to those who know the construction industry, it is a rather broad definition. To sharpen it, consider definitions from several well-respected professionals in the crisis management field.

In *No Surprises,* Mindszenthy, Watson, and Koch (1988) argue that "a crisis can be any threat or event that creates chaos and, usually, suffering" (p. 27).

In *Crisis in Organizations,* Barton (1993), a leader in crisis management planning, offers the following definition: "A crisis is a major, unpredictable event that has potentially negative results" (p. 2).

In *The Crisis Manager,* Lerbinger (1997), a professor at Boston University and a prolific author on crisis management, defines it thus: "A crisis is an event that brings, or has the potential for bringing, an organization into disrepute and imperils its future profitability, growth, and, possibly, its very survival" (p. 4).

In *The Crisis Counselor,* Caponigro (1998) has this definition: "A crisis is any event or activity with the potential to negatively affect the reputation or credibility of a business and is typically a situation that is—or soon could be—out of control" (p. 3).

In *Crisis Management: Planning for the Inevitable,* Fink (1986) has another way to look at the definition of the term. "A crisis is any situation that runs the risk of:

1. Escalating in intensity.
2. Falling under close media or government scrutiny.
3. Interfering with the normal operations of business.
4. Jeopardizing the positive public image presently enjoyed by a company or its officers.
5. Damaging a company's bottom line in any way (pp. 15–16)"*

In the construction industry, a crisis typically means that someone has been injured or killed on a project. This is a realistic definition that is supported by industry statistics. According to the Bureau of Labor Statistics (1999), in 1998 the construction industry suffered a total of 1,171 fatalities. Unfortunately, according to the same report, this fatality statistic increased since 1997, when 1,107 fatalities were reported and the industry had a rate of 9.3 injuries per 100 workers. A sobering fact is that during 1998, the construction industry employed 6% of the U.S. workforce yet was responsible for 19% of workplace fatalities.

Although accidents involving injury or death are those most commonly associated with the construction industry, they are not the only events that can cause a crisis. This book defines a crisis as *any incident that can focus negative attention on a company and have an adverse effect on its overall financial condition, its relationships with its audiences, or its reputation in the marketplace.*

Granted, situations will occur from time to time that will never be known outside the company. In other words, they will remain internal and the outside world will never hear of them. Nevertheless, a host of situations could become known outside the company and potentially spell trouble.

What types of crises could knock on your door?

"Plan for a wide range of small and large disasters," warns Bill Walsh (1993), "and you'll reduce the potential for you and others to be caught off guard. You can actually be more aggressive, and thus be more likely to avert setbacks. Having planned for the worst, you can do a lot more than just hope for the best" (pp. 13–14).† As coach of the San Francisco 49ers, Bill Walsh has had to manage crises similar to those expe-

---

*Copyright Steven Fink, Lexicon Communications Corp., www.crisismanagement.com. Reprinted with permission.

†Reprinted by permission of FORBES ASAP Magazine © Forbes Inc., 1993.

rienced in the construction industry—everything from negative media attention to serious injuries. His warning should be taken seriously by construction managers.

The following potpourri of potential crises is by no means a complete list; however, it does suggest important questions: What else are you vulnerable to, and which crises have the greatest potential of occurring?

Review this list and check off the situations that have the potential to knock at your company's door as well as the ones that have already knocked a hole in it. The purpose of this exercise is to understand that, even with the best safety program, a company is still vulnerable to crisis situations. The key is to identify risks and take the necessary steps to prevent them from turning into crises. Many risks begin subtly but then explode into full-blown crises—for example, a disgruntled employee who later turns to workplace violence. To plead ignorance and assume "It won't happen on my watch" is unacceptable because it courts catastrophe.

Crisis	Could Happen	Been There

Natural Disaster

- Lightning
- Earthquake
- Extreme snow/ice conditions
- Extended freeze
- Flood/drought
- Hurricane/Tornado/Tsunami

Operations

- Equipment failure
- Accident involving a company vehicle
- Bomb threat
- Loss of a key subcontractor/supplier
- Construction delay
- Cost overrun
- Design error/issue
- Explosion
- Fire
- Major utility failure
- Neighborhood/community opposition to a project
- Structural/subsidence collapse
- Data/telecommunications failure/loss of critical data

Crisis	Could Happen	Been There

Environmental Accidents/Liabilities

- Groundwater contamination
- Long-term exposure of the community to toxic chemicals
- Release of toxic chemicals into the air or waterways

Employee Safety and Health

- Chronic safety problems
- Exposure to carcinogens
- Injury/fatality of an employee or nonemployee
- Personal injury suit
- Regulatory citations

Labor Relations

- Negotiations
- Organizing drive
- Unfair labor practices
- Violent strike
- Work stoppage

Management Issues

- Bankruptcy
- Contractual dispute with a client, resulting in litigation
- Employee raiding by a competitor
- Hostage/kidnap and ransom
- Hostile takeover attempt
- Key employee starting a competing company
- Management succession
- Merger/acquisition
- Murder
- Negative publicity from rumors
- Negative publicity relating to political contributions

Crisis	Could Happen	Been There

Management Issues (cont.)

- Death of owner or key employee
- Reorganization/downsizing
- Serious cash-flow problems
- Sudden market shift
- Suicide
- Terrorism
- Blackmail

Employee/Management Misconduct

- Bribery/kickbacks
- Disgruntled employee
- Executive misconduct/fraud/ embezzlement
- Lawsuits from discrimination, sexual/ racial harassment
- Murder
- Price fixing
- Sabotage
- Scandal involving top management
- Slander
- Suicide
- Theft/Vandalism
- Workplace violence

Government Affairs

- Legislation that could affect business

If you have checked even one of these events, you are not alone.

How do you compare with the rest of the industry?

Compare your company with those I surveyed in March 1988 and again in March 1996, including 149 general contractors, heavy/highway contractors, subcontractors, manufacturers, engineers, owners, and construction materials producers. The survey's purpose was to determine the status of crisis preparedness in the construction industry. (See Appendix for full details on who was surveyed.) The participants were presented with a list of 27 crises and were asked to indicate the ones they had experienced in the three years prior to the survey. Here is the Top Ten list for both 1988 and 1996:

1996

#1   On-the-job accident
#2   Damage to utility lines
#3   Contractual dispute with a client, resulting in litigation
#4   Equipment failure
#5   On-the-job fatality
#6   Highway accident
#7   Theft/embezzlement
#8   Noise/dust pollution
#9   Sexual harassment
#10 Labor strike/work stoppage

1988

#1   On-the-job accident
#2   Contractual dispute with a client, resulting in litigation
#3   Damage to utility lines
#4   On-the-job fatality
#5   Theft/embezzlement
#6   Labor strike/work stoppage
#7   Serious cash-flow problems
#8   Rapid growth
#9   Lack of bonding capability
#10 Sudden market shift

As you can see, few changes occurred over the eight-year period between surveys. Highway accidents, noise/dust pollution, and equipment failure were replaced by rapid growth, lack of bonding capability, and serious cash-flow problems.

What was the financial impact of these crises?

Half of the respondents in each survey reported negative impacts on their overall financial picture, both in hard and soft costs. Furthermore, a crisis makes heavy demands on time for both management and the key personnel trying to contain the situation. A company without a crisis management plan becomes a victim of the demands of the crisis and cannot afford the luxury of being proactive. When a company falls into the reactive mode, flailing and floundering, the crisis moves into the driver's seat, hard and soft dollars roll away, and the recovery of those dollars becomes extremely difficult.

EXHIBIT 1.1 Sales Needed to Cover the Direct Costs of an Injury at Various Profit Margins

Direct Cost of Injury	1 Percent	2 Percent	3 Percent
$ 1,000	$ 100,000	$ 50,000	$ 33,000
$ 5,000	$ 500,000	$ 250,000	$ 167,000
$ 10,000	$ 1,000,000	$ 500,000	$ 333,000
$ 25,000	$ 2,500,000	$1,250,000	$ 833,000
$100,000	$10,000,000	$5,000,000	$3,333,333

(*Source:* The Sheet Metal and Air Conditioning Contractors' National Association, Inc. *Industrial Insights,* February 1988)

As mentioned earlier, injuries are prevalent in the construction industry. Now consider the cost. The Sheet Metal and Air Conditioning Contractors' National Association (SMACNA) completed a survey that showed the enormous sales a company needs to generate to pay for the direct and indirect costs of a work site injury. Estimates show that a company operating at a 1% profit margin must make *a million dollars* to pay for the direct costs of a $10,000 accident. Exhibit 1.1 indicates the sales needed to cover the direct costs of an injury at various profit margins.

The indirect costs associated with an injury are considerable. According to Hinze and Applegate (1991), the ratio of indirect to direct costs for medical-case injuries is 4.2 and for restricted activity or lost-workday injuries 20.3 (p. 537). The indirect costs could include a negative swing in employee morale, a decline in productivity, damage to the company's reputation, management time spent on the issue, negative media attention, and so forth. Such soft costs can exceed the hard costs of an accident—yet another reason to identify risks and stop them from becoming crises.

What is the impact on employee productivity?

In my surveys, nearly half of the respondents in 1996 and one third in 1988 noticed a decrease in employee productivity as a result of a crisis.

Depending on the type of crisis, it is not at all unusual for productivity to decline during and after an emergency. The soft-cost effects can range from traumas as a result of witnessing an accident or fatality to negative media attention to a lack of communication between management and employees.

Does a crisis affect a company's reputation?

Among the respondents to my surveys in 1996, 30% felt their reputations were jeopardized or damaged as a result of crises. This percentage is a significant decrease from

1988, when over 75% reported their reputations were in jeopardy. That 41% of the participants had developed crisis management plans by 1996 suggests that they had become increasingly aware of the importance of being proactive and of protecting their most valuable asset—their reputation.

Did the media show up?

In the 1996 survey, 58% of the respondents received media attention as a result of their crises, and more than half felt the coverage was unfair. This percentage is increased from 1988, when 43% received media attention and over one third felt the coverage was unfair. Interestingly, more than half of both groups felt they were treated unfairly by the news media.

The media can be extremely aggressive, especially in a crisis. If reporters believe there is a story, they will not go away until they get it. Every corporate representative must be equipped with the skills necessary to portray his or her company in the best possible light while working under the worst possible circumstances. If a company is not prepared for a media encounter, it will probably get poor coverage—or, worse yet, inaccurate, one-sided coverage. The penalty is severe because there is no practical recourse to set the record straight.

How many of the respondents had a crisis management plan?

Of the 149 respondents in 1996, 41% reported that they had crisis management plans. This figure is encouraging because the 1988 survey revealed that only 23% had plans. Of those who had plans in 1996, 31% developed them as a result of crises. Unfortunately, crises are common; the upside is that if a company develops a plan after a crisis, the next crisis will be handled proactively.

A comparison of the two surveys shows that the number of crises occurring in the industry has not changed. The respondents averaged five crises each over the three years prior to each survey. Does this research imply that the industry is crisis-prone? Or is it time take control of potential risks and no longer accept them as part of doing business?

Can you prevent crises?

In the large majority of cases, the answer is an unqualified yes merely by applying an increased awareness. The only exception might be a natural disaster (an Act of God); however, advanced technology can forewarn us of such impending dangers and allow us to plan for them and the subsequent havoc they create. In man-made crises, a warning bell typically is sounded, but it usually falls upon deaf ears. Here is an example:

Let us suppose that a general contractor is in the structural steel phase of a 15-story commercial office building located in a major metropolitan area. There are 12 sub-contractors on site and a total of approximately 95 employees. It is 7:00 A.M. on a Tuesday morning and a crane is being used to lift steel beams, which are being set by a team of ironworkers. This is all of the information needed for the most senior person on site to ask, "What could go wrong on this project today?" The logical answer might be to focus on the obvious issues related to the equipment; however, a host of situations could ensue:

1. Equipment failure causing structural damage, injury, or death
2. Unsafe safety practices causing injury or death
3. Neighborhood opposition to the project
4. Accident involving a company vehicle
5. Labor unrest
6. Fire/Explosion
7. Chemical spill or hazardous material release
8. Disgruntled employee perpetrating workplace violence
9. Harassment/discrimination issues
10. Complaints to the media about the project
11. Loss of a key subcontractor or supplier
12. Bomb threat
13. Schedule issues affecting costs and the construction company's relationship with the building owner

A project is fluid and changes daily because of the schedule, workforce, equipment, materials, chemicals, and so forth. Therefore, it should become a daily habit for the most senior person on site to ask, "What could go wrong on this job today, and what needs to be done to prevent it?" It is vital for the project team to work closely with safety personnel who are trained in risk identification. Consistent attention to this collaboration is the key to eliminating or minimizing potential crisis situations. This collaboration must begin when dirt is first scratched and not be terminated until the keys are turned over to the owner.

It is understood that the construction industry can be intrusive. Here is an example of how a general contractor identified a potential crisis and turned it into an opportunity for the company:

XYZ Construction was awarded a project for a six-story medical office building in a major metropolitan area. While the project team was celebrating, John, the vice president of operations, reviewed the project for problems that could arise during the construction phase and came up with one that could cause concern.

The project was located directly across the street from a retirement home filled with people with time on their hands. John was fully aware that this project, like any construction project, would cause some disruption and inconvenience to the lives of the residents. So, before any equipment was moved on site, John met with the director of the retirement home and proposed that the director allow him a small corner of the home's lobby to display a model of the finished medical office building and a schedule of activities. He then proposed that he and the project superintendent visit the retirement home every Monday morning at 8:00 A.M. to provide a progress report to the residents and to explain what construction activities would occur that week.

The director thought it was a tremendous idea, and the meetings began. The first meeting was coolly received by the residents, who were not exactly sure they wanted all of this activity in their backyard. John gave all the residents hard hats with the logos of the construction company and the retirement home on them and made each resident an honorary project manager. He also gave the residents the superintendent's phone number to call if they were uncomfortable or did not understand an activity at the project site. Because the project was located in a high-crime area, John also asked them to call the company if they saw any unusual activity during off-hours.

The result? Over 80% of the residents attended the meetings and became involved with the construction process. They also became the contractor's security company, advising XYZ Construction of kids trying to access the site during off-hours or putting graffiti on the sidewalk barricades. At the topping-out ceremony, the local paper did a front-page story showing the retirement home residents in front of the project with their hard hats on, clapping and cheering.

This is a classic example of how a company can anticipate a potential problem and prevent it from turning into a crisis. John averted a potential community and public relations crisis and turned the situation into a fabulous opportunity that garnered goodwill for both his company and the owner of the project.

The identification and prevention of risks falls on the shoulders of upper management, whose finger must be on the pulse of the company at all times and whose awareness of risks must be keen either to avoid them or to minimize the damage they can cause.

Can all crises be identified? Although there may be surprises, for the most part, all members of a project team as well as upper management should be so intimate with their respective areas of responsibilities that surprises should be minimal.

How can a crisis go from bad to worse?

Every crisis is unique and takes on a life of its own; however, the key to effective crisis management is anticipation. A chain reaction will most likely occur as a result of a crisis, and spin-off crises can create more damage than the initial crisis itself. When a problem hits the fan, one must be able to look beyond the initial crisis and determine the chain of events that could occur. This foresight can be accomplished through a process called a *what-if analysis*.

What is a "what-if" analysis?

In a what-if analysis, you assume the role of a negative creative thinker and take a look at the domino effect that could occur as a result of a crisis. Creatively, you envision the events and actions that could occur in conjunction with an identified crisis. Then you ask yourself, "What else could happen?" Remember, it is not necessarily the initial crisis that deals the crippling blow to your company; it could very well be an event that occurs as a result of the initial crisis and that your company could have identified and avoided.

To anticipate all of the scenarios, you construct a what-if diagram, which takes the form of a branched tree and which on each branch documents the possible occurrence of an additional crisis. In developing the various scenarios, the main question to ask is, "What is the worst that can happen?" The what-if diagram becomes a graphic presentation of a possible crisis progression. Because of its graphic nature, this diagram should help to track each of the many conceivable crisis developments.

This exercise is best accomplished in a noncrisis environment, when all possible spin-off crises can be anticipated without panic. Exhibit 1.2 shows an example of a what-if diagram that resulted from such a brainstorming session.

Now let's change gears and track the what-if process with a crisis in progress. Our hypothetical situation involves a construction materials company that has just initiated a blast. Everything went off as planned except for one small piece of flyrock. The direction the flyrock took was thought to be toward a remote area, so no one believed that this flyrock would be a cause for alarm. Work proceeded as normal and the flyrock was not thought about again until an hour later, when the plant manager received a phone call. It seems that the flyrock found its way through a residential roof and into someone's living room.

During the ensuing chaos, someone decides to predict the ramifications of this incident via what-if analysis. The thought process may look something like this:

> What if a resident is injured? What if a resident is killed? What if friends and relatives of the injured make statements critical of the company? What if exaggerated damage claims are filed? What if the residents complain to the media? What if the company is deluged with calls from the media? What if the media show up at our plant, demanding answers to questions? What if the media print and broadcast stories critical of our safety procedures? What if political officials call with questions?

As you can see, a what-if analysis can ask a number of provocative questions, but its true goal is to bring critical thinking into the situation to give you a chance to be ahead of the crisis instead of the target of criticism. The what-if process is crucial and should be implemented immediately upon notification of an incident so that spin-off crises can be anticipated and eliminated or minimized.

EXHIBIT 1.2 What-if Analysis

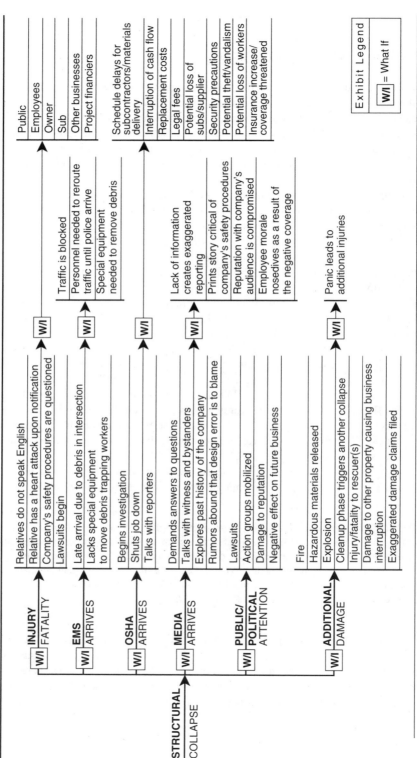

(*Source:* Janine Reid Group, Inc.)

What is the impact of your crisis on your company and various audiences?

First, let's define *audience*. The word refers to anyone who can have an effect on your business or reputation. This effect might include, but is not limited to, the following groups:

Shareholders/investors
Stock analysts
Board of directors
Employees and their families
City/county officials
Site neighbors
Opinion leaders
Action groups
The media—both the general press and the trade press
Governmental regulators
Financial community
Clients (current, past, potential)
Other contractors, architects, engineers
Unions (if applicable)
Suppliers
Insurance/bonding company

The effects of your crisis can be far-reaching. Decisions, which are made rapidly during a crisis, can have a tremendous effect on your audiences. As illustrated in Exhibit 1.3, several questions need to be addressed when a crisis strikes:

What effect will this crisis and our actions have on these areas:

- Our operations?
- The general public?
- Our industry?
- Our financiers?

In the heat of a crisis, there is not a lot of time for a committee to convene and discuss the pros and cons of each of these questions. The crisis management team must be able to get its arms around the situation quickly and have its actions reflect the culture and values of the company.

Communicating with your various audiences is vitally important, especially if the news media becomes involved. This issue is discussed at length in Chapter 8; for

EXHIBIT 1.3 21 Crisis Impact Questions

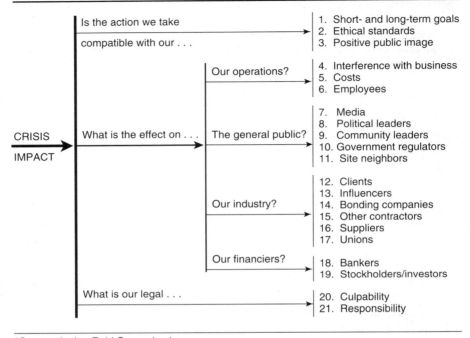

(*Source:* Janine Reid Group, Inc.)

now, rest assured that all of your audiences will be paying very close attention to your actions and the impact that they will have on them. As the late Louis R. Perini of the Perini Corporation once said, "A good reputation, once acquired, is never permanent."

Chapter Summary

- A crisis is any incident that can focus negative attention on a company and have an adverse effect on its overall financial condition, its relationships with its various audiences, or its reputation in the marketplace.

- There is no substitute for a solid safety program that is supported and endorsed by upper management. However, even with the best safety program, crises can still occur. A proactive company takes the time to educate its employees on the importance of identifying risks and preventing them from becoming crises.

- Upper management should challenge themselves and their key employees to identify each risk and then take measures to prevent its becoming a crisis. This precaution requires only a few minutes of a manager's time, yet it can save a company untold dollars and stress.

- Once a crisis occurs, a domino effect will likely ensue. Apply a what-if analysis to determine other events that could occur as a result of the initial crisis. Simultaneously, think about conducting an impact analysis to determine which of the company's audiences are affected by the event. Organization and communication are critical areas in crisis management planning.

2

THE CRISIS MANAGEMENT TEAM

Why is a team needed?

Having worked in a variety of crisis situations, I have found that it is virtually impossible for *one* person to handle all of the tasks necessary to bring a crisis to a close effectively. No matter what the size or scope of the crisis, it will require a variety of people with different areas of expertise to ensure that all facets of the problem are covered. Therefore, a team approach to crisis management is essential.

Whether you are an owner/developer, program manager, construction manager, engineer, architect, general contractor, or subcontractor, your crisis management team should be a permanent unit that can draw on internal and external resources. The objective is to have a knowledgeable group that works effectively and quickly in a crisis at both the corporate and the project level.

What are the selection criteria for team members?

The crisis management team should comprise individuals who not only have complementary areas of knowledge but also are willing to work in a team environment under stressful conditions. Team members must be able to make decisions rapidly because a crisis will not afford them the luxury of spending hours around a conference table discussing the pros and cons of each step and *then* taking a vote. Time is not on the side of the crisis management team; and the stress created by the accelerated pace, coupled with the information vacuum at the outset of the crisis, may prove intolerable for many people.

Understanding how individuals respond under pressure is critical to the selection of team members. When the unimaginable becomes reality, behaviors change. The stress of working on a crisis, especially if a human life is threatened can precipitate emotions and reactions that normally do not surface in the day-to-day work environment. Some of those reactions may include the following anomalies:

- Typically high-performing individuals may have reduced performances. The intense level of stress created by a crisis can paralyze the decision-making ability of some people and so cause them to become part of the problem.
- Because the company's normal chain of command will most likely vanish when the crisis team is mobilized, some people on the team or in the company may feel discomfort both during and after the crisis. When the team is activated, the company's hierarchy very likely will change abruptly and perhaps radically. Subordinates may suddenly have authority over superiors, and the crisis management team itself may mix people from several levels of authority. Such organizational shifts may become part of the problem too.
- Problematic personality traits may become exaggerated from the tension of a crisis. Normally high-strung individuals may become even more so, aggressive behavior may be exacerbated; and timid personalities may withdraw totally. Such altered personalities may become part of the problem as well.

In *Crisis Management: Planning for the Inevitable,* Fink (1986) offers an excellent metaphor to express the pressure that the crisis management team can expect: "It is like walking through a maze backward, wearing a blindfold, and juggling lighted sticks of dynamite" (p. 144). Although team members' responses to stressful situations will not be fully tested until a crisis occurs, advanced crisis planning and training will increase the team's chances for success. So will the selection of backup personnel for key positions on the team. If someone becomes too stressed out, a trained substitute should be at the ready.

Who are the members of the core crisis management team?

It is recommended that you select your core crisis management team based on the largest potential crisis that could befall your company. This statement usually provokes the question, "What if I run a lean company and have only four people on my salaried payroll? I simply don't have enough people to fill a crisis management team!"

Good point. Nevertheless, it is important to understand that crises do not discriminate between small and large companies, and they do not care if you have 4 or 600 hundred salaried employees. Certain objectives need to be accomplished when a crisis hits; and whether you use internal or external sources, those objectives do not change because of the size of your company.

Your team should be built envisioning the most catastrophic crisis that could happen to your company. It should consist of employees and/or outside consultants who not only possess their respective areas of knowledge but who also are willing to work on the team.

Let's start with the selection of your core crisis management team. The selection of core team members is based on two important factors:

1. Each member should possess an area of expertise that is useful to the team. For example, each member must have good organizational capabilities, have the trust and confidence of upper management, possess solid communications skills, and have the respect of the company's employees.
2. The team members should work well together in normal day-to-day situations. If they do not, whatever problems exist among them will be exacerbated in a crisis.

There are four members of your core crisis management team:

1. Team leader
2. President
3. Legal counsel
4. Spokesperson

These four members will be involved in any event classified as a crisis and will mobilize other team members as the situation warrants. Exhibit 2.1 shows the core crisis management team, along with support personnel and possible outside help that may be activated. Because the chain of command during a crisis may alter the company's usual organizational chart, all such changes must be clarified during the selection process *before* any crisis occurs.

What are the responsibilities of the core crisis management team?

There are two major goals to shoot for when a crisis occurs: *organize internally* to get the situation under control as quickly as possible and *communicate externally* your side of the story and what you are doing about the incident. While all team members are vital to the process, two important team members are selected to ensure that these goals are achieved. They are the team leader and the spokesperson.

Team leader

The team leader is the company's *internal organizer* during a crisis, selecting and mobilizing a team to control the crisis as rapidly as possible with the fewest complications. All team members report to the team leader, who should be a member of middle to upper management and have a working knowledge of the company's short- and

EXHIBIT 2.1 Core Crisis Management Team

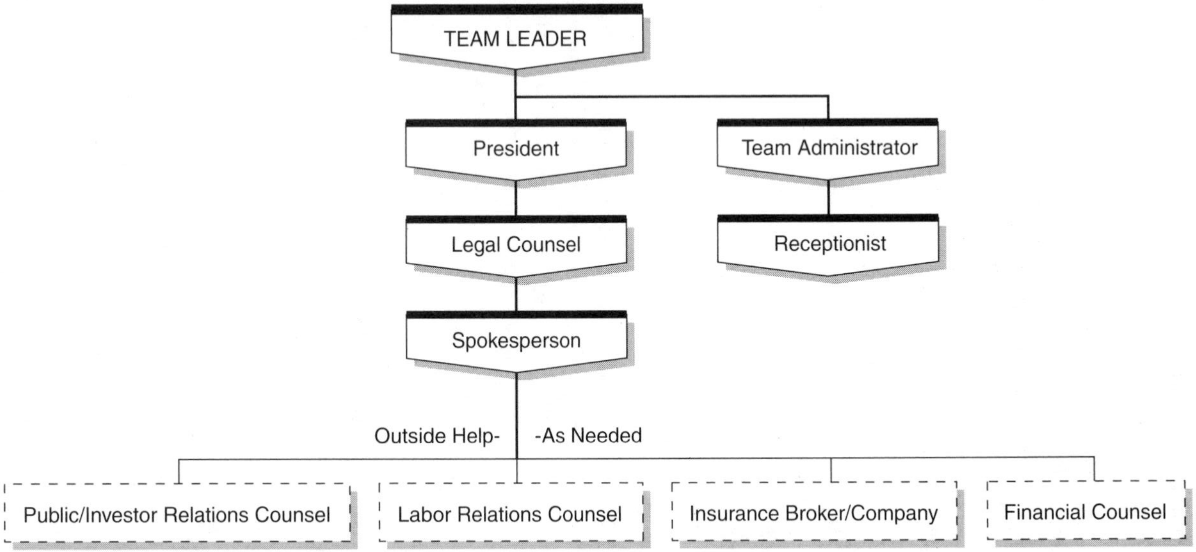

TEAM LEADER

President

Legal Counsel

Spokesperson

Team Administrator

Receptionist

Outside Help-   -As Needed

Public/Investor Relations Counsel

Labor Relations Counsel

Insurance Broker/Company

Financial Counsel

(*Source:* Janine Reid Group, Inc.)

long-term goals. Making rapid decisions is the norm for the team leader and requires that he or she have access to and the confidence of the company's president and other top managers. Here are the desirable characteristics a team leader:

- Strong organizational skills
- Ability to motivate and work effectively with all of the company's managers
- Assertiveness, but not aggressiveness
- Strong problem-solving capabilities
- Willingness to commit to what could be a very large responsibility
- Willingness to replace team members who cannot handle the stress of the event and prove a detriment to the team

The company's president should not take on the role of the team leader because a crisis makes heavy demands on time and because he or she must be available for other duties described later in this chapter.

As you can imagine, few individuals have all of the personality and professional traits to fulfill the role of the team leader. Look for as many of these desired characteristics as possible because the importance of selecting the right team leader cannot be overemphasized. A wrong choice will probably show itself at the worst time— namely, during a crisis. The responsibilities of the team leader are as follows:

1. Assess the situation quickly by asking the five famous questions of Who? What? When? Where? and How?
2. Mobilize a team based on the answers to the above questions.
3. Notify and work closely with upper management.
4. Advise the team administrator (described later on) and the receptionist on how to route calls.
5. Determine what spin-off crisis could occur by using the what-if analysis technique described in Chapter 1.
6. Communicate with all employees via fax, e-mail, and/or voice mail if the crisis is getting outside attention. Tell them the way they should handle requests for information and the person to whom the requests should be referred.
7. Notify the human resources department, if the situation calls for it.
8. Notify the insurance broker/company, if the situation warrants.

Select a backup team leader just in case the primary team leader is on vacation, at a remote job site, or otherwise unavailable. Many companies have two to three back-ups to make sure the coverage is there all the time.

Spokesperson

The spokesperson is the company's *external communicator* to its various audiences. As defined in Chapter 1, a company's audiences consist of shareholders/investors, board of directors, employees and their families, political leaders, site neighbors, opinion leaders, actions groups, governmental regulators, city/county officials, the financial community, clients, other contractors/architects/engineers, unions, suppliers, insurance/bonding companies, and so forth. While each of these audiences must be contacted during a crisis, the primary responsibility of the spokesperson is working with the news media, should they become involved in the crisis.

Working with the news media is an area where companies seem to get into trouble if they are not prepared. In today's business arena, any company that does not take the time to understand the media, learn its needs, and become comfortable with them probably will get poor coverage or inaccurate, one-sided coverage. The penalty is severe because there is no practical recourse to set the record straight. This penalty drives home the importance of selecting and training a spokesperson *before* your company finds itself in the spotlight. If you wait until a crisis occurs to make your selection, the results could be less than favorable. Your spokesperson should be able

- To handle a high level of stress and be unflappable
- To think quickly on the spot
- To neutralize situations and maximize opportunities
- To be articulate and provide honest information without creating legal issues or further problems
- To take direction from the team leader and upper management

The best candidate for the position of spokesperson is an individual who already speaks to reporters or outside groups on behalf of your firm. If you do not have a qualified candidate in your firm, now is the time to develop and train one. The spokesperson has three major responsibilities:

1. Give the media the highest priority during a crisis. Communication with reporters must be swift, consistent, and controlled in terms of content.
2. Work closely with the team leader and upper management in developing a communication strategy.
3. Document all statements given to the media. This documentation should include the name of the reporter, publication/station, date, time, and information delivered. This documentation will be described in more detail in Chapter 3.

Additional responsibilities may be assigned in the event of a job-site accident; these are outlined in Chapter 5. As with the team leader, the president should not be placed in the position of the spokesperson. There are several reasons why:

- It is difficult for a president to say, "I don't know." Reporters expect a person in this position to know everything and be prepared to share it.
- The president simply will not have the time to be the media liaison in a crisis because of other responsibilities that require his or her attention.

- The president can possess too much knowledge about the incident or past incidents and so may read more into a reporter's question than necessary.
- What if the president makes a serious mistake while giving an interview? There is no one in a higher position to correct the error.

Despite these considerations, there are two exceptions to this rule:

1. A tragic event may require an initial response from the president, acknowledging the incident, communicating concern, and delivering an update on the status of what is being done about the situation. This statement could be delivered at the scene of the accident or at the news conference. Chapter 7 elaborates on this subject. After that point, the responsibility of media relations should be handed off to another member of the crisis management team.
2. If the company has limited resources and the president is the *only* candidate to fill the role of spokesperson, then so be it.

As with the team leader, a backup spokesperson is highly recommended not only to take on the responsibility if the primary spokesperson is out of town but also to relieve the primary spokesperson if necessary. During a major crisis, for instance, the primary spokesperson can become overwhelmed and so require some relief. Also, legal counsel should *not* be used as a spokesperson because placing a lawyer in that position may be interpreted as a sign of guilt. Legal counsel may also have a tendency to withhold information in an effort to protect the client; however, because the court of public opinion can be unforgiving, the message intended by a lawyer may not be the message received by the media and then communicated to your audiences and the general public.

Companies with regional or district offices should have a core crisis management team at each location. This applies to remote project locations as well.

President

The president, or another company official, must make it his or her responsibility to allocate time to follow the crisis through to its conclusion. The responsibilities for the president are:

- Work closely with the team leader and legal counsel to determine the company's direction and position on the incident.
- Review and approve all statements released to the media.
- Determine which of the company's audiences are affected by the incident and ensure that communication with these audiences is established and maintained throughout, and beyond, the crisis.
- Personally notify the next of kin in the event of a fatality of a company employee. This notification is discussed in more detail in Chapter 3.

- Be prepared to make the *initial* statement to the news media in the event of a tragic incident, to show concern and personal involvement.
- Approve the crisis management plan and all additions or changes that occur during periodic updates.

Legal counsel

As part of your core crisis management team, your legal counsel should have a full understanding of your crisis management plan and the external communication demands that are part and parcel of managing a crisis. From an internal perspective, your legal counsel needs be contacted at the first indication of a crisis and kept apprised of all decisions during a crisis because legal guidance could prove invaluable. Your legal counsel, whether in-house or on retainer, should also have a current copy of your crisis management plan and be familiar with the players on your team.

From an external perspective, it is common, at the outset of a crisis, for your legal counsel to request that you have no external communications with the news media due to possible admission of guilt or liability reasons. However, this is a difficult request to accommodate; Chapters 4, 5, and 6 offer suggestions on how to handle such situations.

As mentioned, there are two supporting members of your core crisis management team—the team administrator and the receptionist. The following details their responsibilities.

Team administrator

The team administrator is the right hand of the team leader and a critical support member of the crisis team. The responsibilities of the team administrator are as follows:

- Provide support to the crisis team—for example, assisting in the screening of phone calls, organizing strategy meetings, maintaining communication with team members, arranging for support services for the family or families of the injured, dispatching a critical-incident stress counselor to witnesses and personnel, and performing clerical tasks.
- Update and distribute changes and additions to the crisis management plan on a quarterly or as-needed basis.
- Work with the receptionist on how to route calls during a crisis.

Receptionist

The person handling the phones during a crisis is critical, yet is typically the last one to know what to do. Example: the corporate headquarters of a large general contractor received word that an explosion occurred on one of the company's job sites. Needless to say, word spread quickly around the office and people started to panic. A reporter heard about the incident over a scanner, called the main office, and asked the receptionist what was happening. Her response was, "I don't know what's going on except that all hell is breaking loose!" You can take it from there.

It is the responsibility of the team leader or the team administrator to contact the receptionist at the first warning of an impending crisis and to provide the receptionist with the following directives:

- Route all requests from the media to the person screening calls for the spokesperson.
- Never release information to outside callers, no matter how hard you are pushed.
- Do not allow a reporter to roam about the office unescorted; always notify the spokesperson of a reporter's presence as soon as possible.

Exhibit 2.1 lists outside counselors who may be called upon, depending on the type of crisis. Their responsibilities are defined in the next section.

Who are the members of the crisis management team for a project?

From a project standpoint, the crisis management team should be selected prior to job startup; however, it is never too late to develop a team. There should never be uncertainty about who is in control should a crisis occur. A good rule to follow is that *the highest entity on the project has the responsibility to select the crisis management team for the project* and then communicate the assigned responsibilities to the respective parties. This selection is crucial because organization is paramount in a crisis, so individuals need to know who is in control. To do otherwise results in total chaos.

First, define who is in charge in the event of a project-related crisis. Exhibit 2.2 shows the entities that could be involved in a project. The highest entity among them who reports to the project owner/developer should be in charge in the event of a crisis. That rule does not absolve the other entities on the project from planning for a crisis. Every company involved should have a protocol to follow in the event of a crisis, but at the outset of the project it must be determined who has ultimate control.

Now, consider a crisis management team in action. As previously discussed, it is important to establish a team based on the largest and most complex crisis that could potentially happen to your company and, in this case, on one of your projects. Exhibit 2.3 illustrates a crisis management team for a $100-million general contractor. Members of this team are selected and mobilized based on the size and scope of the crisis. Potential members of the project crisis management team are as follows:

Team leader
President
Spokesperson
Technical spokesperson

Project owner/developer
Team administrator
Receptionist
Project manager
Superintendent
Foreman
Temporary spokesperson
Safety manager
Human resources
Legal counsel
Insurance broker/company
Outside help as needed

You may be thinking that this seems like a lot of people for a crisis, but keep in mind that not every crisis will require the participation of the full team. Only those with the necessary expertise need be mobilized.

We have reviewed the responsibilities of the core crisis management team, so let's build on that team as illustrated in Exhibit 2.3.

EXHIBIT 2.2 Who Is in Charge?

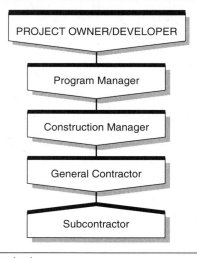

(*Source:* Janine Reid Group, Inc.)

EXHIBIT 2.3 Project Crisis Management Team

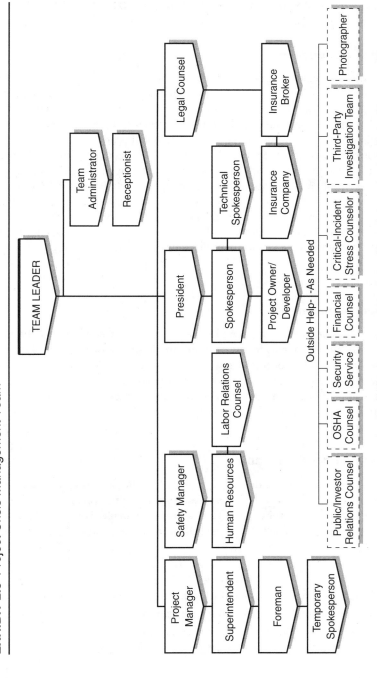

(*Source*: Janine Reid Group, Inc.)

27

What are the responsibilities of the crisis management team for a project?

Project owner/developer

Many companies feel that approaching project owners/developers on the issue of crisis management may send a signal that the companies are crisis-prone. I have found the contrary to be true. Project owners/developers are receptive to anything that will protect their reputations and relationships in a crisis, and they look upon companies with a crisis management plan as being proactive and progressive.

A project owner/developer should be notified immediately of an incident and become involved to the level at which they are comfortable. As you can imagine, this involvement could present interesting communication challenges if a protocol is not established in advance of the crisis. Many public entities, such as a port authority or a department of transportation, have their own public information officers, and they may wish to be the team leaders or spokespersons during crises. Advanced planning can help to eliminate many of the complicated communication issues that are certain to arise.

Project manager

Most small- to mid-sized projects do not have a project manager on site. In the event of a crisis, the project manager may be called to the site, if the situation warrants, and assist the superintendent or foreman with his or her duties (defined below). If on site, the project manager works with the superintendent/foreman to ensure that the crisis management plan is put into effect. Also, the project manager most probably acts as the temporary spokesperson until the corporate spokesperson arrives.

Superintendent/foreman

In the event of a job-site accident, the project superintendent (or foreman, if there is no superintendent on the project) is most likely the senior person on site. In most cases, this person has total control of the site. If a project manager is on site, he/she works with the superintendent/foreman on the following action items. Many of these items are discussed in Chapter 3.

- Contact and coordinate whatever emergency services are necessary.
- Secure the site and determine if it should be shut down.
- Contact the team leader and safety manager and request their assistance.
- Make certain that any and all evidence remains intact and unmoved.
- Work with subcontractors to obtain a head count and identify witnesses.
- Arrange for the witnesses to be debriefed.

- Determine, with the assistance of the team leader, who will notify the next of kin in the event of an employee injury or fatality and arrange for bilingual communication if needed.
- Post workers to restrict entry to the site until it is deemed safe.
- Notify the owner/developer of the project.
- Instruct the job-site receptionist on how to route phone calls.
- If a temporary spokesperson is needed, work with the team leader to determine the best candidate.

Temporary spokesperson

If a crisis occurs on a job site and the media arrive, the odds are slim that the corporate spokesperson can get to the site in time to be of immediate assistance. If such a delay is the case, you need a temporary spokesperson to buy time with reporters until the primary spokesperson can be mobilized. This temporary spokesperson is necessary to show the media that the company is not stonewalling. A "buy-time" statement consists merely of acknowledging the incident, stating what is currently being done, and guiding the reporters to a safe area to await further information from the temporary or primary spokesperson. Chapter 6 defines this process in greater detail.

So the question is, "Who should be the temporary spokesperson?" The answer is that the senior person on site must either take on this responsibility or assign it to another member of the project team. Because of the chaos created by the incident, the last thing the senior person on site wants to do is to talk to a reporter. The natural inclination is to say "No comment" to reporters and kick them off of the site. As tempting as this thought is, you must resist the urge and understand not only that you are in control of the job site but also that you are in control of the dissemination of information.

Technical spokesperson

Construction is a highly technical arena and can be misunderstood by nonexperts. At issue during your crisis might be a complicated design or a piece of equipment that failed. If so, you may need to call on an expert in a specialized area who can explain such matters and answer technical questions. This spokesperson usually enters the scene well after the crisis is underway—perhaps in the second or third day of the crisis. You must have supreme confidence in your technical spokesperson because good reporters find their own experts who can, potentially, undermine yours.

Safety manager

It is the safety manager's responsibility to assist the project manager or superintendent/foreman in securing the site and ensuring that further risks are eliminated. Additional responsibilities are as follows:

- Notify the necessary authorities.
- Begin the debriefing process with witnesses and contact a critical-incident stress counselor (described later in this section) if necessary.
- Determine if the company should initiate a third-party investigation team to work with the authorities.
- Document the incident both in writing and on film.
- Act as liaison with the medical facilities, should the incident involve an injury or fatality.

Human resources/labor relations counsel

The human resources department is the gatekeeper to the company's personnel records, policies, and emergency contact numbers. A representative from human resources should be available during a crisis to handle the following action items:

- Provide employee information in the event of an injury or fatality.
- Begin working with the insurance company, if needed.
- Act as the conduit to the company's employees. When a crisis occurs, it is imperative to tell the staff what is happening and what to tell the outside world—if asked. Employees are constantly talking to suppliers, friends, subcontractors, and so forth, and it does not take long for bad news to travel. Therefore, human resources should use current employee communications systems to notify all employees about the crisis, the company's actions in handling it, the statements the employees should make about the situation, and the person to whom they should refer all media calls. This procedure also applies to branch offices and remote job sites.
- Work with the labor relations counsel should a union issue arise.

Insurance broker/company

If you feel that your insurance program will be affected by the incident, contact your claims adjuster as quickly as possible. The claims adjuster will most likely proceed to your site to help secure the evidence and begin the insurance company's investigation.

Outside help

A crisis may require expertise that is not available within your company. Exhibit 2.3 outlines possible resources for your consideration.

Public/investor relations counsel

From a public relations standpoint, a crisis is any incident that focuses unusual media attention on your company. Obviously, crisis communications should be planned before there is a crisis, when it is possible to create a reservoir of goodwill. This step is discussed further in Chapter 9. For now, remember that no matter how complete your plan, you have failed if you do not communicate your message to your audiences. A reputable public relations or investor relations firm or consultant with expertise in crisis communications can be an invaluable resource during a crisis. Here are some traits to look for when searching for such counsel:

- Previous experience in crisis communications for either private or publicly held companies.
- Working knowledge of the local media and, if needed, the regional and national media.
- Ability to interview people, synthesize their responses, and write statements and releases that position the company in its most favorable light.
- Access to technology to get the company's message out to its various audiences quickly via Web site, e-mail, video news release, wire services, etc.
- Ability to act as the buffer between the media and the company spokesperson— in other words, to screen the spokesperson's calls and provide counsel on how to answer questions. It is not recommended, however, that your public relations counsel take over the role of corporate spokesperson for two important reasons:

  1. This counsel does not work for the company and therefore cannot answer most of the questions.
  2. Reporters want to talk to someone in authority and so have little patience with an intermediary. They will see right through the screen of using a public relations person.

As you can see, working with the media during a crisis can be frustrating. Another set of eyes and ears experienced in this area can be helpful in providing guidance in this process.

OSHA counsel

If the Occupational Safety and Health Administration (OSHA), the Mine Safety and Health Administration (MSHA), the Environmental Protection Agency (EPA), or another entity is conducting an investigation, the safety manager may elect to secure the services of a consultant who has a strong working knowledge of all relevant regulations and can provide the following assistance:

- Facilitate communications between OSHA and the company.
- Assure that the company is complying with OSHA requests in a timely manner.
- Protect the company's interests throughout the investigation.

Security service

Some situations require the assistance of a security company. Such situations include, but are not limited to, labor unrest, bomb threats, violence issues, and accidents. It is useful to explore potential companies before the need arises. Otherwise, in the press of events during a crisis, you will probably revert to the largest ad in the yellow pages, and that choice may not be the best.

Financial counsel

Depending on the type and severity of the crisis, the controller (or whomever you select) should be available to advise the team leader of the availability of service/ product suppliers or cash resources. For example, a helicopter may be needed to remove debris from an intersection, the company president may need to charter a plane to get to a distant location, or the family of an injured employee may be in need of cash until they can secure their own resources. The company's financial counsel should be able to respond to these needs on very short notice.

Critical-incident stress counselor

As discussed in Chapter 1, the construction industry experiences a high rate of injuries and fatalities as well as a plague of other misfortunes. Witnesses of these events traditionally have been expected to return to work as quickly as possible without the benefit of counseling because of an old attitude that says, "Accidents are just part of the job." It is past the time for the construction industry to rectify this ancient attitude. In *Critical Incident Stress Debriefing,* Mitchell and Everly (1996) warn against this attitude:

> Despite the fact that Critical Incident Stress Management has been written about and practiced for almost 20 years, there are some people who still believe that it is either not needed or that only those who are weak or otherwise badly suited for their jobs are in need of support services. This false belief has caused some administrators and supervisors to actively fight the development of traumatic stress management or, for that matter, any stress management programs within their organizations. The cost of this short sighted and badly thought out management strategy may never be adequately measured. Since many people suffer though the painful situations they encounter in silence, no one may ever know just how many people have prematurely left their jobs, suffered deterioration in their health, endured negative changes in their relationships or personalities, or lost the joy of living and working in their chosen careers as a result of unresolved work place stress (p. 57).

Ten years ago, I witnessed a tragic accident and fell victim to the attitude that I did not need "help." The following 90 days were a blur of mixed emotions and reactions to that decision. I had periodic bouts of recurring flashbacks, wild emotional swings, interrupted sleep from nightmares, decreased productivity, anger, and depression.

I am now keenly aware of the importance of critical-incident stress counseling and strongly recommend that companies contact a trained counselor to debrief their witnesses and project team as soon as possible after an incident (within 24 to 72 hours). Even if the project team did not witness an incident, feelings of guilt and responsibility need to be dealt with. Mitchell and Everly (1996) claim that "critical incident stress debriefing accelerates the rate of normal recovery in normal people who are having normal reactions to abnormal events" (p. 15).

What about employees who do not directly witness the accident? Well, a crisis has a rippling effect and touches all of your employees. Many companies have, with great success, brought in a critical-incident stress counselor to speak to their employees about the incident and some of the residual effects they might expect. This counseling brings a great deal of comfort to the company as well as facilitates an internal support system that can be relied on throughout the crisis.

Third-party investigation team

In addition to one or more authorities investigating a situation, a company may elect to call in third-party counsel to conduct an independent study of the incident. The purpose of this study is to provide the company with an evaluation independent of both its own and the investigating authority's evaluations.

Photographer

Many crises require film documentation of evidence for investigation purposes, legal reasons, or simply for the project file. Pictures should be taken from every possible angle and have a time and date stamp on each image. It is recommended that a professional photographer be retained to ensure high-quality images as well as verification— in court, if necessary—of the photographs. Film documentation should begin at the outset of the investigation and conclude at the direction of the team leader.

The typical medium used is 35mm; however, there may be a call for instant results, which requires the use of a Polaroid camera. A digital camera can be extremely useful to e-mail the images to the corporate office or other locations, if necessary. If you choose not to secure the services of a professional photographer, assign this responsibility to a crisis management team member.

How do you track the team members' activities?

A crisis demands fast communication and even faster decisions. The team leader must be able to rely on the instincts and decision-making capabilities of the team. In return, the team members must provide the team leader with oral updates of their activities at least twice a day and a written report at the conclusion of each day. These daily reports by all team members serve as logs of their activities, detailing the people they communicated with, the things said or promised, the progress made in their respective areas of responsibility, and the items that need to be covered the next day. This daily report serves a number of additional purposes:

1. Events move very quickly in a crisis, and the discipline of a daily report allows team members the time to reflect on and record their activities and formulate plans for the next day.

2. If backups are called in to assist, they can become current on the situation by reviewing the daily reports.

3. The daily report can assist legal counsel as well as serve as a postemergency evaluation tool, which is discussed in Chapter 11.

This discipline of completing a daily report must be maintained throughout the crisis. This information is crucial from an organizational and communication standpoint because the goal of crisis management is to be ahead of, not behind, the rush of events.

Chapter Summary

- No matter what the size or scope of the crisis, it requires a variety of people with different areas of expertise to ensure that all of the crisis management plan is put into effect. It is virtually impossible for one person to handle all of the tasks necessary to bring a crisis to a close.

- The two most important members of the crisis management team are the team leader and the spokesperson. The team leader is the company's internal organizer and the spokesperson is the company's external communicator. Other team members must be chosen carefully; personalities and professional abilities must be assessed as accurately as possible. The team members make the difference between a solution and another crisis.

- Always assign a backup to each of the core crisis management team members. Crises have a habit of occurring during the night or on weekends when team members are unavailable or cannot be reached. Backups can also serve as a relief to the primary members to reduce fatigue, which is a natural side effect of stress caused by trying to manage a crisis.

- Exhibit 2.2 indicates the chain of command in a crisis situation. The highest entity reporting to the project owner/developer is in charge in the event of a crisis.

- Develop your own crisis management team based on the largest potential crisis your company could have. Consider all of the possible outside resources that may be called upon and establish a relationship with them before the crisis hits the fan.

3

COMPONENTS OF A CRISIS MANAGEMENT PLAN

Why is a plan so important?

Ugly headlines—loss of business—tarnished reputation—poor employee morale—these are a few of the negative effects of a poorly handled crisis.

A crisis management plan is an integral component of a company's safety and risk management program and should be part of strategic planning. The goal of a crisis management plan is to reduce the potential adverse effects of a crisis by ensuring an organized response to them. It also provides the crisis management team with a blueprint for handling crises before one bursts on the scene.

Experience has shown me that a company's audiences judge its worth by the way it handles a crisis even more than by the crisis itself. If a company is not armed with a crisis management plan, it may appear inept and poorly managed during a crisis. I have had the opportunity to work on dozens upon dozens of construction-related crises, and there is absolutely no comparison between a company that has a written plan and has trained employees on its use and a company that has not. The company that has included a crisis management plan as part of its risk and insurance management program stands a greater chance of maintaining control during a crisis, retaining customer and employee loyalty, minimizing its financial losses, and quickly restoring its operations than the company without a plan. The prepared company can react more quickly with an increased ability to think through action items from a number of angles. The company that does not have a protocol to follow is compromised in such matters and may fall victim to half-baked decisions.

However, a crisis management plan is not a panacea, nor is it a guarantee that no one will point an accusatory finger in your direction. Nevertheless, it takes the panic out of the situation and gives you the presence of mind to be proactive rather than reactive during the rush of events. "Nothing narrows your options faster than a crisis" is a proverb in the construction industry, but a plan allows you to think through your options before they are needed.

The goal of developing an effective crisis management plan is to complete the thinking process and detail work under calm conditions instead of the duress of a crisis. Because no two companies are alike, this chapter presents many ideas that can serve as guidelines for customizing your specific plan. These ideas are offered to stimulate your critical thinking skills so you can make the best out of a difficult situation. Your job is to determine what should be included in your specific plan and what modifications are necessary to serve your particular needs.

Every crisis is unique; however, common patterns surface in many crises. Instead of trying to develop a plan for every conceivable crisis, such as those identified in Chapter 1, consider the commonalties of all of those crises and develop your plan around them.

What should be included in your corporate plan?

The secret to an effective crisis management plan—that will *actually be used* during a crisis—is to keep it simple. Let me repeat—keep it simple! No one has time to read a philosophical thesis when the world is crashing down around them.

With that in mind, I present the following menu of items for your consideration for inclusion in your corporate crisis management plan:

Cover sheet
Control sheet
Table of contents
First-hour response checklist
Contacts in the event of an emergency
Salaried personnel contacts list
Notification procedures in the event of a serious employee injury
Notification procedures in the event of an employee fatality
Procedures for identified crises
Community relations
Tips on working with the news media
Safety history
Personal safety for international travel
Past emergencies
Company fact sheet
Project data sheets
Foreign language expertise

Emergency procedures contact list
Incident report
Evacuation plan

Next, I describe each component's content and note whether I feel that the item's inclusion is mandatory or optional. If an item is noted as mandatory, you should know that many years of field experience with crises have taught me the value of that item. If an item is indicated as optional, either it may not be applicable to your firm or it is meant to be used as support data throughout the duration of a crisis. The components of this plan can be adapted to any type of firm. You need change only a few words to have it apply to your company. Again, the following ideas are presented as suggestions only, and their selection for your plan is up to you.

Cover sheet—mandatory

The purpose of a cover sheet is to indicate the type of plan within—that is, corporate, district, regional, project, plant, and so forth. The plan should also be marked as confidential because home phone numbers of the crisis management team are included in it. Proprietary information should not be included in the plan—just in case a copy wanders outside of your doors.

The bottom of the cover sheet should indicate the person the plan was assigned to as well as the date of the next revision. The team administrator should review the master plan a minimum of once a quarter. In short, the cover sheet is mandatory because it indicates who the plan was developed for, who it has been assigned to, what the next revision date is, and what kind of plan it is.

Control sheet—mandatory

The team administrator distributes and updates the company's crisis management plan to select individuals at various locations. This distribution is discussed later in this chapter. The control sheet is mandatory because the team administrator must maintain a log sheet to track the name and title of each person assigned a plan along with the location of his or her plan for update and revision purposes.

Table of contents—mandatory

Your reader needs to have not only a thorough understanding of what is included in your plan but also a way to find information in it quickly. A table of contents provides such a quick reference. Next, there should be a one- to two-paragraph executive summary stating the purpose of the plan. Then a list of sections should follow, with a brief description of each section's contents. The sections of your plan should be tabulated and labeled for ease of access.

First-hour response checklist—mandatory

Ask anyone who has lived through a crisis what the first hour was like, and you will probably hear that it was total chaos and that no one could think clearly. The first hour of a crisis is somewhat analogous to a paramilitary operation. A company must have directives to ensure that all of the targeted items are covered by the appropriate people.

The first-hour response checklist is a critically important item in your crisis management plan because it guides the crisis management team through the action items that need to be covered immediately upon notification of a crisis. Exhibit 3.1 is a checklist developed for a general contractor for a job-site accident involving injuries

EXHIBIT 3.1 First-Hour Response Checklist

Step One—Senior Person on Site

_____ Contact emergency services.
_____ Contact the safety manager.
_____ Determine if the site should be shut down.
_____ Make certain that all employees are accounted for.
_____ Do not move anything that could be classified as evidence.
_____ Ensure telephone coverage at the site.
_____ Tell job-site personnel where to direct information requests.
_____ Notify the crisis management team leader.
_____ Post workers to restrict entry to the site until it is deemed safe.
_____ Select a temporary spokesperson with the assistance of the team leader and deliver a buy-time statement.
_____ Notify the owner/developer of the project.

Step Two—Team Leader

_____ Determine what happened, when and where it happened, and who is involved.
_____ Determine who is investigating the emergency.
_____ Verify the current status of the job site—is it shut down?
_____ Decide whether the team leader and/or spokesperson is needed on site.
_____ Notify management.
_____ Advise the team administrator and the receptionist how to route calls.
_____ Identify potential spin-off crises.
_____ Fax/e-mail/voice mail all employees and job sites to notify them of the incident and advise them who will handle media and general information calls.
_____ Notify human resources.
_____ Notify the insurance broker/company.

Step Three—Safety Manager

_____ Gather the names of the injured and/or deceased.
_____ Obtain the phone number(s) of the spouse(s)/family(ies).
_____ Contact the team leader to determine who should notify the spouse(s)/family(ies).
_____ Debrief workers who witnessed the accident.
_____ If necessary, initiate a postaccident drug/alcohol test (check with legal counsel).
_____ If appropriate, notify the applicable governmental agency.
_____ Determine if a third-party investigation team is needed.
_____ Designate someone to stay with the injured worker(s) at the hospital until family members arrive.
_____ Document the incident in writing and on film.

EXHIBIT 3.1 Continued

Step Four—Team Leader

____ For injury/fatality, notify the appropriate union (if applicable).

____ If there is an employee injury/fatality, determine who will notify the spouse/family.

____ If the injury/fatality is a subcontractor's employee, it is the subcontractor's responsibility to notify the spouse/family.

____ If a nonemployee is hurt/killed, allow the authorities to make the notification and contact your insurance broker/company.

____ Inform any surrounding areas that may be affected by the incident.

____ Instruct employees to contact their families to let them know they are OK.

Step Five—Spokesperson

____ Write, and get clearance for, all statements and releases.

____ Designate someone to screen your calls from the news media.

____ Complete the media log sheets.

____ Anticipate media questions. If possible, role play a media interview with a colleague before going live.

____ Assemble the necessary background information and literature.

____ If you elect to give the media a tour, make certain that the area is safe and that they are escorted by a company representative. Issue safety equipment and require that a hold-harmless agreement be signed.

____ Instruct reporters on your safety procedures before going on site. If they violate any of the procedures, you have the right to ask them to leave.

____ Advise reporters of a time and place for future updates.

____ Follow up on additional media inquiries.

Step Six—Team Leader/Human Resources

____ Identify the audiences that need to be contacted for update purposes.

____ Gather details on past negative issues to which the media may refer.

____ Track all media coverage via a monitoring service and the Internet.

____ Establish communications with all employees and provide ongoing updates.

____ Secure and offer critical-incident stress counseling for employees who witnessed the accident.

____ Complete a post-emergency evaluation at the conclusion of the crisis.

(*Source:* Janine Reid Group, Inc.)

and/or fatalities. For this example, let's say that the general contractor is the entity in charge (as illustrated in Exhibit 2.2) and reports directly to the project owner. Now let's walk through each action item and justify its existence.

Step One—Senior Person On Site

Contact Emergency Services. As far as calling 911 is concerned, every jurisdiction is different, so do your homework prior to a crisis. Some jurisdictions operate a combined center for paramedics, fire, and police. When you place your call, the operator asks you about the type of emergency you are experiencing and then forwards your call to the appropriate agency for dispatch. If you are using a cell phone, be sure to give the dispatcher your phone number and the address at which the crisis is occurring.

Because the news media monitors police and fire department dispatches, do not be surprised if reporters appear either before or shortly after emergency services arrive. In an effort to delay this inevitable visit by the media, some companies elect to contact bypass 911 and call the needed service directly. Experience has shown me that a crisis is not the time to play the system. If you need assistance, you typically need it fast and your employees will remember 911 better than a seven- or nine-digit number.

Many companies find it useful to invite emergency management services to their sites for tours before emergency services are needed. This practice is especially recommended for projects that are remote or hard to find.

Contact the Safety Manager. The safety manager needs to be sent to the crisis scene as quickly as possible to provide assistance in securing the site, to ensure that further risks are identified and eliminated, to assist in the investigation, and to fulfill the other duties described in Chapter 2.

Determine If the Site Should Be Shut Down. The senior person on site, in coordination with the authorities, makes this determination based on the severity and location of the crisis. If it is a very large job site, the determination may be made to shut down the area where the incident occurred but continue operations on the rest of the site.

Make Certain That All Employees Are Accounted For. Immediately following the accident, the general contractor must initiate a head count of all employees. The senior person on site should have each subcontractor's superintendent/foreman submit a head count of his or her crew as quickly as possible. At this time, all witnesses should be identified and escorted to the job-site trailer for debriefing.

Do Not Move Anything That Could Be Classified as Evidence. The senior person on site must make sure that all evidence remains intact until the investigating parties have approved its removal.

Ensure Telephone Coverage at the Site. Because the phone at a job site experiencing a crisis is extremely busy, a receptionist must be assigned to handle calls, take messages, and direct calls to the appropriate people. The person assigned this responsibility is not authorized to disseminate information related to the crisis. Instead, the receptionist's sole purpose is to make sure that messages are getting to the appropriate people and that the crisis management team can remain in communication with the site.

Tell Job-Site Personnel Where to Direct Information Requests. The accident may attract attention from outside entities such as neighborhood groups, environmental groups, the media, bystanders, and so forth. These individuals may have access to your employees, subcontractors, and suppliers if your project is not secured by gates or fences. As the senior person on site, it is your responsibility to make certain that your company speaks with one voice. Thus, you must tell all employees not to release information about the incident but rather to refer inquiries to you or whomever you designate.

Notify the Crisis Management Team Leader. The preceding seven action items must be taken care of immediately upon notification of a crisis. The next step is to contact the team leader at the corporate office and ask for his or her assistance. Verifiable information must be relayed and support requested where needed. Then the team leader must mobilize a team to provide assistance, both at the site and at headquarters.

Post Workers to Restrict Entry to the Site Until It Is Deemed Safe. The last thing you need is to have outsiders on site asking questions and subjecting themselves to potential dangers. From a safety perspective, the site must be secured as quickly as possible.

Select a Temporary Spokesperson with the Assistance of the Team Leader. If the incident is likely to draw media attention and the primary spokesperson cannot get there in time, a temporary spokesperson must be selected. This position may be filled by the senior person on site or delegated to someone else. The decision should be based on who can do the best job rather than who has the highest title. If the crisis is related to a subcontractor, the subcontractor should refer all media calls to the general contractor's temporary spokesperson. Sample buy-time statements for the temporary spokespersons are detailed in Chapter 5.

Notify the Owner/Developer of the Project. The project owner/developer needs to be notified as soon as possible after the crisis has occurred. As discussed in the previous chapter, the project owner/developer may elect to participate on the crisis management team, so his or her notification is vital. Whether participating on the team or not, the owner/developer still needs to be notified as soon as possible. No one enjoys discovering on television, on the radio, or in the paper that his or her project is in crisis.

Step Two—Team Leader

When a crisis erupts, the senior person on site is overwhelmed with action items and has little time to spend on the phone. As team leader, you must respond to his or her call quickly and gather as much information as possible so you can mobilize a crisis management team. First, determine the answers to the following questions:

- What happened, when and where did it happen, and who is involved?
- Who is investigating the emergency?
- What is the current status (is the job site shut down)?
- Is the team leader and/or spokesperson needed on site?

Notify Management. After gathering the above information, relay it to the president or another member of management immediately. If you are the team leader in a district office, you must notify the top in command at your location as well as the management team at the corporate office.

Advise the Team Administrator and Receptionist How to Route Calls. Keeping track of phone inquiries is a monumental task and must be handled with extreme care. Remember, the receptionist answering the phones does not have the authority to release information about the crisis. Rather, the receptionist's responsibility is to route calls to the individuals selected by the team leader.

Identify Potential Spin-Off Crises. The team leader, or whoever the team leader designates, must apply the what-if analysis (see Exhibit 1.2) to the situation. The goal is to identify as many potential spin-off crises as possible and do whatever is necessary to prevent their occurrence.

Fax/e-Mail/Voice Mail All Employees and Job Sites to Notify Them of the Incident and Advise Them Who Is Handling Media and General Information Calls. Because the rumor mill works fast inside and outside the company during a crisis, immediately notify employees of the incident and of the steps the company is taking to deal with it. Also, ask that they refer all media calls to the company spokesperson and any other questions or concerns to the team leader.

Notify Human Resources. Human resources can assist with employee communications as well as provide employee information in the event of an injury or fatality.

Notify the Company's Insurance Broker/Company. Your claims adjuster will proceed to the site as quickly as possible and begin an investigation.

Step Three—Safety Manager

Immediately on arrival, the safety manager must determine who has been injured, the extent of the injuries, the medical facility receiving the victim(s), and the phone number(s) of the spouse(s)/family(ies). The safety manager then contacts the team leader to determine who should notify the spouse(s) and/or family(ies) of the injured. (This procedure is described in detail later in this chapter.) If the injury/fatality involves a subcontractor's employee, the subcontractor must notify the spouse(s)/family(ies); however, the general contractor must make certain that this action takes place. In some cases, the authorities make the family notification, but the injured employee's company must make certain that it has made contact as well. If the injury or fatality involves a union employee, the craft steward or business agent may fulfill the notification responsibility.

Debrief Workers Who Witnessed the Accident. Witnesses should be taken to an area away from the incident location, perhaps the job-site trailer, and individually asked to recount the accident. The senior person on site determines who is responsible for interviewing and documenting each witness's recollection of the event. This information should be gathered as soon as possible after the incident because the more time passes, the more the story may change. The written documentation is sent to the corporate team leader who, in turn, submits it to legal counsel.

Witnesses should then be referred to a critical-incident stress counselor. This referral is described later in this section.

If Necessary, Initiate a Postaccident Drug/Alcohol Test. Consult your legal counsel and state laws about the ability to request postaccident drug/alcohol testing.

If Appropriate, Notify the Applicable Governmental Agency. Depending on the type of work you are doing as well as the type of crisis you are experiencing, the decision to proactively contact OSHA, MSHA, or the EPA is the responsibility of your upper-management team. Each crisis is different, so the rules and regulations of each agency must be closely examined.

Determine If a Third-Party Investigation Team Is Needed. As discussed in Chapter 2, a third-party investigation team may be called in to initiate an investigation independent of the company or authorities.

Designate Someone to Stay with the Injured Worker(s) at the Hospital Until Family Members Arrive. An individual should be assigned to meet the spouse/family at the medical facility and remain as long as necessary. Do not abandon or lose touch with those injured and their families. Make certain that they have a contact name and phone number of someone who can help them through this process and provide any support they may need.

Document the Incident in Writing and on Film. Self-explanatory.

Step Four—Team Leader

If there is an employee injury/fatality, determine who should notify the spouse/family. (A detailed review of the notification process appears later in this chapter.) If the injury/fatality is a subcontractor's employee, the subcontractor must notify the spouse/family; the senior person on site must make sure that this notification happens in a timely manner. If a nonemployee is hurt or killed, allow the authorities to make the notification and contact your insurance broker/company. Again, this process must be completed as quickly as possible.

Inform surrounding areas that may be affected by incidents connected with crises such as fire, explosion, releases, debris obstructing roadways, outages, and so forth. Coordinate with emergency services if an evacuation is required.

Instruct unharmed employees to contact their families to let them know they are safe. If the incident is receiving media attention, family and friends inquiring about their loved ones may overload your phones. Anticipating this reaction can provide relief to your receptionist as well as curtail the rumor mill.

Step Five—Spokesperson

Write, and Get Clearance for, All Statements and Releases. The team leader and president must clear all communications from the company to the outside world.

Designate Someone to Screen Your Calls from the News Media. The spokesperson should avoid taking a media phone call cold. It is far more advantageous to have a receptionist screen the call and gather pertinent information required for the media log sheet, which is explained in Chapter 5. Then the receptionist should advise the reporter that the spokesperson will return the call as quickly as possible. This procedure allows the spokesperson a few minutes to collect thoughts and position a statement based on the reporter requesting the information.

Complete the Media Log Sheets. Chapter 5, Exhibit 5.3, reviews media log sheets in detail.

Anticipate Media Questions. If Possible, Role-Play a Media Interview with a Colleague Before Going "Live." The spokesperson must apply a what-if analysis just as the team leader does. This analysis answers the question, "If a reporter asks me this question, what should I say?" Such anticipation reduces the level of stress in an interview.

Assemble the Necessary Background Information and Literature. Reporters may request information on your company as well as the project where the accident happened. The crisis management plan should provide much of that information.

If You Elect to Give the Media a Tour, Make Certain That the Area Is Safe and That They Are Escorted by a Company Representative. A tour is allowed only when the site has been deemed safe by the safety manager and any authorities on site. If the incident draws a large number of reporters, you can restrict your tour to one reporter and one photographer only. The media then decides who will be sent and will pool all of the information gathered. Never allow anyone on site who is not fully attired with the proper safety equipment and who has not signed a hold-harmless agreement, which is a release of liability.

Instruct Reporters on Your Safety Procedures Before Going on Site. Reporters have been known to wander into off-limits areas during a post accident tour to take pictures and ask questions. This is trespassing. If reporters violate your directives, you have full authority to ask them to leave—and then make certain they do so. If problems arise, ask the police to take care of the situation for you. Safety is your highest priority, and if you are the enforcing entity, it is your job to make certain that the site is safe.

Advise Reporters of a Time and Place for Future Updates. Once you deliver a statement, reporters are anxious to know when further information will be available, so you should advise them of the time and place of the next information update.

Follow Up on Additional Media Inquiries. Inquiries from the media should always be answered in a timely basis. Doing otherwise results in coverage such as "The contractor could not be reached for comment" or "The contractor had no comment." Such language implies that you have something to hide; consequently, your audiences may receive an unfavorable impression of your company.

Step Six—Team Leader/Human Resources

Identify the Audiences That Need to Be Contacted for Update Purposes. As discussed in Chapter 1, a company must quickly identify which of its audiences may be affected by the situation and communicate its side of the story to them.

Gather Details on Past Negative Issues to Which the Media May Refer. Today's technology affords the media the luxury of conducting an investigation into your past within a very short time—especially if this incident is similar to a previous occurrence. This topic is discussed later in this chapter.

Track All Media Coverage via a Monitoring Service and the Internet. It is extremely important to know what is being reported by the news media. Assign a service or individual the responsibility of monitoring television, radio, and newspaper coverage throughout the crisis. This monitoring helps you anticipate future media questions and gives you an opportunity to correct inaccurate information.

Establish Communications with All Employees and Provide Ongoing Updates. In a crisis, employees can become either loyal supporters and help the company send the right messages to its audiences, or they can be the source of negative statements and leaks. To avoid the latter, a company must communicate quickly and honestly with its entire employee base, and the message must be consistent with the one presented to the media and to the company's other audiences. However, even though management may make a strong effort to update employees and tell them to refer media calls to the spokesperson, a disgruntled employee may still use a crisis as an opportunity to air grievances. If so, those grievances must be monitored and a response developed in case someone takes the disgruntled employee seriously.

Secure and Offer Critical-Incident Stress Counseling for Employees Who Witnessed the Accident. The emotional impact of a crisis can be a heavy burden for anyone, especially those who witnessed the incident. Witnesses may be apprehensive about returning to work and fear that they can become victims themselves. Such feelings can be surprisingly powerful and may interfere with productivity. Thus, as stated in Chapter 2, offering critical-incident stress counseling cannot be overemphasized.

Complete a Postemergency Evaluation at the Conclusion of the Crisis. When the worst of the crisis is over, it is the team leader's responsibility to review the performance of the crisis management team as well as conduct an evaluation of the crisis management plan. Revisions should be made at that time.

Again, this checklist was prepared for a general contractor to cover a job-site accident involving an injury or a fatality. A checklist can be modified for any type of crisis; however, as discussed in Chapter 2, your checklist should be developed around the largest potential crisis that could befall your company.

Contacts in the event of an emergency—mandatory

When a crisis occurs, a large number of people, companies, and agencies need to be contacted and mobilized. Time is precious during a crisis, and it can be needlessly wasted tracking down contacts and phone numbers. The purpose of the plan is to have this information organized for you, prior to a crisis, in an easy-to-use format. Exhibit 3.2 offers an example for your review. Develop your list based on the contacts your company may need in a time of crisis and group them by function. For instance, one group of contacts and numbers may be the members of the crisis management team, the next, various consultants, the next, various emergency services, and so forth.

EXHIBIT 3.2 Contacts in the Event of an Emergency (Add pager numbers if applicable)

Crisis Management Team	Telephone Numbers
Team Leader	Day: Night: Cell:
Backup	Day: Night: Cell:
Spokesperson	Day: Night: Cell:
Backup	Day: Night: Cell:
President	Day: Night: Cell:
Vice President(s)	Day: Night: Cell:
Safety Director	Day: Night: Cell:
Safety Consultant	Day: Night: Cell:
Human Resources	Day: Night: Cell:
Team Administrator	Day: Night: Cell:

EXHIBIT 3.2 Continued

Counsel

Corporate Counsel	Day: Night: Cell:
OSHA Counsel	Day: Night: Cell:
MSHA Counsel	Day: Night: Cell:
EPA Counsel	Day: Night: Cell:
Labor Relations	Day: Night: Cell:

Insurance and Related Services

Insurance Company	Day: Night: Cell:
Insurance Broker	Day: Night: Cell:
Workers Compensation	Day: Night:
Employee Assistance Program	Day: Night:

Consultants

Crisis Consultant	Day: Night: Cell:
Workplace Violence	Day: Night: Cell:
Kidnap/Ransom	Day: Night: Cell:
Public/Investor Relations	Day: Night:
Photographer	Day: Night:

continued

EXHIBIT 3.2 Continued

Emergency Services

EMS	911
Hospitals	See Project Data Sheet
Urgent Care Clinics	See Project Data Sheet
Occupational Health Clinics	See Project Data Sheet
Ambulance	911
Police	911
Fire Department	911
Poison Information Center	
American Red Cross	

Utility Companies

Power Company
Gas Company
Phone Company
Water Company
Traffic Signal Repair

Governmental Agencies/Offices

OSHA
MSHA
EPA
HazMat
Local Health Department
Coast Guard
County Officials
Mayor's Office
Governor's Office
Chamber of Commerce
Building Codes

Transportation Services

Taxi Services
Airplane Charter
Helicopter Charter

Alternate Office Location

(In the event your corporate office is inoperable)

(*Source:* Janine Reid Group, Inc.)

Corporate media list—optional

It is useful to have a list of phone numbers for television stations, newspapers, and popular radio talk shows in the geographic areas where you have a presence. This list makes calls for updates and clarification of information easier. If you have offices located in other cities and states, it is useful to have a media list that is geared to those locations as well. Here are directories that can assist you in the development of your media list:

Burelle's Media Directory
Burrelle's Information Services
75 East Northfield Road
Livingston, NJ 97939

Bacon's Directory
Bacon's Information, Inc.
332 South Michigan Avenue
Chicago, IL 60604

News Media Yellow Book
Leadership Directories, Inc.
104 Fifth Avenue
New York, NY 10011

Salaried personnel contact list—optional

If a crisis occurs in the evening or on a weekend, you may need to contact your employees either to notify them of the situation or to ask for their assistance. I have indicated that this list is an optional item because many companies do not like to distribute home phone numbers of their salaried employees.

Notification procedure in the event of a serious employee injury—mandatory

Notification is a high-priority item and should occur as quickly as possible. Notification via telephone is acceptable in an injury situation. The goal is to get the family/spouse to the medical facility as quickly and safely as possible and let the attending physician release the status of the injured. It is recommended that a cab or car be dispatched to the spouse or family member's house/place of business prior to the notification call. An individual receiving such a call is in a highly charged emotional state and could create another crisis. Exhibit 3.3 offers a checklist for notification of the employee's spouse/family in a serious injury situation.

Notification procedure in the event of an employee fatality—mandatory

A personal, face-to-face notification of the employee's spouse/family is mandatory if an employee of the company is killed. A telephone call is unacceptable. The only exception to this rule is if the spouse/family is in another city/state and a company official cannot get to the location in a timely manner. In that case, a local authority, such as a minister or police officer, should be contacted to make the personal notification,

---

EXHIBIT 3.3 In the Event of a Serious Employee Injury

1. Determine the extent and nature of the injuries.

2. Find out immediately where the person is being taken.

3. **The senior person on site and the team leader** determine the most appropriate person to call the spouse/family. That individual explains that there has been an accident and that the employee has been injured, but does not discuss the severity of the injuries. If the spouse/family asks about the severity of the injuries, the response should be: "We can't be certain of the extent of the injuries until we hear from a doctor."

   Advise the spouse/family that a cab is arriving momentarily to take them to the medical facility. Discourage them from driving themselves unless they absolutely insist.

4. If necessary, send an employee to the injured employee's house to lend assistance. This help may include offering a ride to the hospital (if a cab was not used) or finding someone to watch the children (if applicable).

5. The team leader assigns someone to stay in contact with the hospital to monitor the injured person's condition.

NOTE: If the injury involves a nonemployee, the authorities should be consulted about notification procedures. Also, contact your insurance company and legal counsel as soon as possible.

(*Source:* Janine Reid Group, Inc.)

---

and then the company should follow up. There is absolutely no excuse for not making this notification as quickly as possible. Exhibit 3.4 provides a checklist for the notification process in the event of an employee fatality.

Notifying a family member of a fatality is a difficult job for anyone. The emotional impact of the wide range of emotions and reactions can push the messenger to the

---

EXHIBIT 3.4 In the Event of an Employee Fatality

1. **A member of the company's upper-management team** makes a "best effort" to inform the spouse/family *in person* of the accident. If it is not possible to make a face-to-face notification, a minister or police officer may be a possible candidate. The goal is to notify the family quickly—a phone call is a last resort because of its impersonal nature.

2. The designated company representative remains at the employee's home until other family members arrive or for as long as he or she can.

3. The media may attempt to contact family members. You cannot prevent them from talking to the media. It is their right to speak to the media if they wish.

4. Determine whether the employee's family is in need of money to cover small expenses. If so, it may be appropriate to provide assistance in this area. The few dollars spent will come back in goodwill.

5. Maintain contact with a relative or close friend of the spouse or family to ensure that funeral arrangements and related items are being handled.

NOTE: If the fatality involves a nonemployee, the authorities should be consulted about notification procedures. Also, contact your insurance company and legal counsel as soon as possible.

(*Source:* Janine Reid Group, Inc.)

limit. For this reason, it is important not carry out this responsibility alone—*always* take someone along for support and assistance. After the notification, it is important that the messenger seek counseling because the burden of this responsibility will not disappear quickly. Exhibit 3.5 offers guidelines to help the messenger through this difficult task.

EXHIBIT 3.5 Fatality Notification

In the event of an employee fatality, you may be called upon to notify the spouse or family member. This is a traumatic event for both the relative and you. Here are guidelines to help with this process.

- **Do your homework.** Obtain the full name, address, and social security number of the deceased. Then get the full name of the next of kin, the relationship (wife, brother, mother, etc.), and determine if the family members speak English. Find out if the family member has health problems that could be exacerbated upon notification. If so, bring a health-care professional with you. Gather all information relative to the case so you can provide an explanation.
- **Determine where you will meet.** Will the contact be at home, work, or school? If it is outside of the home, arrange with the relative's employer or school for a private place to meet. Verify that you are talking to the correct person—for example, ask, "Are you Sandy Johnson's sister?"
- **Do not go alone.** Take a fellow employee, a friend of the deceased, a member of the clergy, or a police or fire official to support you.
- **Decide in advance what you will say.** There is no easy way to say that someone has died, so speak simply and directly. Terms like "mortally wounded" only confuse people. While it is not necessary to be blunt or cold, at some point it is necessary to say "dead" or "died." Example: "Mrs. Jones, there was a very bad accident this morning at the project. Charlie was moving a ladder and fell over a guardrail. The paramedics did everything they could, but he died instantly."
- **Do not lie.** If you tell a mother that her son died with her name on his lips but she later learns his death was immediate, there is a conflict. It may not be necessary to offer all of the details. Example: If the spouse asks, "Did he suffer much?" an appropriate answer might be, "I don't think so."
- **Be prepared for emotions.** There will be shock, denial, grief, numbness, and anger. These emotional reactions will be directed at the deceased, at you, and at the medical staff. Let the relative vent these feelings. Use common sense and do what seems appropriate at this time. Some people appreciate the touch of a hand; others do not.
- **Decide what not to say.** By not preparing what to say, you may end up saying things that you later regret. Example: In an effort to offer words of comfort, do not say, "He's with God now" or "You're young and will find someone else." Instead, say, "I'm so sorry this has happened to you" or "What can I do to help you right now?"
- **Always listen.** The formula is 90% listening and 10% talking. If the relative needs to go to the hospital or funeral home, you may offer to drive or get a cab. If children are involved, help arrange for a sitter or have a friend to look after them. When appropriate, offer assistance in getting in touch with the life insurance company, social security, and so forth.
- **When it is over, take care of yourself.** You have gone through an extremely stressful event. Take care of yourself now. Use your critical-incident stress counselor to review the difficult process you went through. No one ever gets comfortable with this part of the job.

(*Source:* Curtis H. Childress, CSP, ALCM, The St. Paul Construction)

Procedures for identified crises—mandatory

In Chapter 2, you identified the crises to which you could be vulnerable. I do not recommend that detailed documentation be completed for each identified crisis; however, I do recommend that a first-hour response checklist be developed to assist your team through the critical first hour of a crisis. The exceptions to this advice are crises such as a fire at your corporate office or job site, a bomb threat called into your corporate office or job site, and a hazardous material spill at a job site. For these crises, evacuation routes, search techniques, and containment procedures should be documented and included in either your crisis management plan or your environmental safety and health plan.

For those companies working in third-world countries, the threat of kidnapping is real. You should work closely with your legal counsel and insurance company to develop a kidnapping and ransom plan and keep it separate from all other plans. This plan is highly confidential, so only a limited number in upper management should know how and where to access this information, should it be needed.

Community relations—optional

Let's face it—construction is an intrusive industry. Even though the final product may enhance the surrounding community, the process of construction can be uncomfortable for those around it with safety and health concerns.

The story in Chapter 2 about the general contractor working with the retirement home residents is an excellent example of planning for a construction project to make allies rather than opponents in the neighborhood. Planning and sensitivity can have a strong impact on building positive community relations. The lack of planning and communication in this area can create a very uncomfortable relationship for the duration of your project and, if left unchecked, can shut down the project. As Margaret Meade stated, "Never doubt that a small group of thoughtful and committed citizens can change the world. Indeed, it is the only thing that ever has."

Exhibit 3.6 gives tips on fostering a proactive relationship with the local community through a public meeting or gathering. This community relations item is marked as optional because the type of work your company does dictates whether or not it is needed.

Tips on working with the news media—mandatory

This is a mandatory section of your crisis management plan because a company in crisis must be proactive with the news media in order to tell its side of the story. This section should include sample statements, questions to anticipate and potential responses to them, tips and techniques on working with the news media, and the way to track your responses and the resulting coverage of them. Chapters 4, 5, and 6 offer detailed examples of interactions with the media for your review and selection.

Safety history—mandatory

Your corporate safety program may be challenged when a crisis hits; therefore, it is important to communicate to all of your audiences what your safety program includes.

EXHIBIT 3.6 Community Relations

Solid communication between a company and the surrounding community is vitally important to the success of a project. The following tips are offered to promote a proactive relationship at a public meeting or other gathering.

Come Prepared

Understand the issues and concerns prior to the meeting. Your lack of preparation will be apparent very quickly. Prepare visuals to support your key points and make them large enough for easy viewing. Anticipate all of the potential questions that could arise and role-play your response with a colleague. The importance of rehearsal cannot be overemphasized.

Listen

Community meetings can get out of control because of sensitive and emotional issues. It is important to remain calm and politely listen to the concerns being expressed.

Respond

When the opposition has run its course, restate two to three points that were made—preferably the points you are prepared to address. This step will communicate that you were listening to their concerns and that you are prepared to address them.

Live Up to Your Promises

If you say you are going to do something, make certain that it is accomplished in the stated time frame. If you promise to keep the community apprised of your progress, start a database and mail residents a letter or newsletter at timely intervals.

Research Your Audience

Rapport can be built simply by doing homework on your audience—for example, how long they have lived in the area and their key concerns. This research can tear down the us versus them wall and allow for a healthier dialogue.

Illustrate Your Company's Involvement

Communicate the different ways your company supports the community, such as the number of local residents your company employs, organizations and community efforts your employees are involved with, your contribution to the local tax base, and so forth.

Isolate the Troublemakers

No doubt there will be some people who would like to share their many grievances. If someone starts to dominate the forum, calmly acknowledge that you have heard their concerns and that you would like to talk with them, one on one, after the meeting to reach a resolution. At that point, quickly ask for another question from the other side of the room.

(*Source:* Janine Reid Group, Inc.)

This section in your crisis management plan should cover the high points of your safety program as well as any awards or recognition that your company has received for safety. Once again, this section needs to be completed before a crisis occurs, while calm heads prevail. Exhibit 3.7 provides a starting point for you to customize your safety history sheet.

EXHIBIT 3.7 Safety History

XYZ Construction Company is very concerned about job-site safety and has a well-established, comprehensive safety and health program. We have received numerous awards for safety and for many years our incident rate has been well below the national average. We have published an Environmental Safety and Health Policy and Procedures Manual that is distributed and administered on all of our projects.

XYZ Construction Company requires weekly safety meetings at each job site for its employees, and attendance is mandatory. We also request that the senior employees from all of our subcontractors attend or that they hold their own meetings.

Our safety manager, [name], is dedicated to instilling the importance of safety on all of our jobs with all of our employees.

Safety banners and safety posters are displayed at all job sites.

XYZ Construction Company meets or exceeds all local, regional, and national safety standards and recommendations.

Awards	Year

(*Source:* Janine Reid Group, Inc.)

Personal safety for international travel—optional

If your company is working internationally, your plan should include ways for employees to avoid personal crises while they are traveling. Exhibit 3.8 provides guidelines and Web site information.

Past emergencies—mandatory

If your crisis is receiving media attention, reporters will most likely investigate your history for other negative events. This investigation can be accomplished by searching a database or by simply talking with your employees, subcontractors, or suppliers. Therefore, research the last ten years of your business and identify previous negative events that may surface during a current crisis. You can conduct your own research via the internet by exploring the archives section of newspapers and publication Web sites. More extensive database searches can be accomplished via your legal counsel. I recommend a Nexis search on your company because Nexis seems to be the most comprehensive database for such searches. It provides extensive coverage of newspapers, magazines, and trade journals as well as transcripts of some radio and television broadcasts. Using your company name, your legal counsel can do a word search

EXHIBIT 3.8 Personal Safety for International Travel

Below are tips for avoiding a personal crisis while traveling internationally.

Travel

- Check with your travel planner and the U.S. State Department for travel safety warnings relative to your destination.
- Check the *Website* http://travel.state.gov/travel_warnings.html for Consular Information Sheets. You will find information about the location of the U.S. embassy in the country you are traveling to, currency information, immigration practices, health conditions, entry regulations, crime and security information, and more.

Health

- Check the *Website* http://www.cdc.gov/travel/travel.html for health conditions and specific vaccinations required for the country you are traveling to. You can also call 800-232-1311.
- If you take prescription medications, make certain that you pack an extra supply in the event that your trip is delayed or that you are unable to find a pharmacy that can accommodate you. Pack your meds in your carry-on luggage in the event your baggage gets lost.

Transportation

- Avoid taking taxis. Arrange to be met at the airport and ask the person meeting you to wear his or her company name prominently so that you make the identification, not the other way around. Hotel limousines are considered safer than taxis.
- If you rent a car, check it carefully for damage and make sure everything is working before you leave the lot.
- If you are traveling to a country where kidnapping is common, make certain that you vary your routine every day. Leave at a different time and take a different route to your destination, etc.

Hotel

- Always use every lock available on your door. If you arrive late at your hotel, ask a guard to escort you to your room.
- Do not carry large sums of money. Rely on traveler's checks and credit cards. Keep a list of your credit card numbers in a place other than your wallet in the event your wallet is stolen.

(*Source:* Janine Reid Group, Inc.)

of these sources in Nexis. If a negative article or transcript surfaces, your legal counsel can send you a copy. At that time, use the copy to develop a one-paragraph report that states what happened, who was involved, what the outcome was, and what you learned from the experience. It is better to be prepared to position past crises than to be ambushed and deliver a disjointed response.

Company fact sheet—mandatory

A company fact sheet is a succinct story of your firm's background, location(s), type of work, and community involvement. Simply put, it is an executive summary of your corporate brochure without all of the advertising hype. If the news media become involved in your crisis, they will most probably want to obtain information about your company. Once again, this is information that you should write *before* the pressure of a crisis is upon you. Exhibit 3.9 provides a menu of possible inclusions on your company fact sheet.

The last two items on the company fact sheet show the media and the general public what your company has accomplished professionally as well as its contribution to the community. Include professional awards your company has received or other accolades to show the company's professional work and corporate citizenship.

Project data sheets—mandatory

A data sheet should be completed for each project in progress (over three months in duration) and included in the corporate crisis management plan. Exhibit 3.10 suggests items for inclusion in your project data sheets.

The purpose of the project data sheet is to minimize the search time for pertinent information on a project in the event of a crisis.

EXHIBIT 3.9 Company Fact Sheet

Number of employees:

Corporate office location:

Location of other offices:

Geography served:

Services offered:

Annual dollar volume:

Key management:

Business/community involvement:

AWARDS:

(*Source:* Janine Reid Group, Inc.)

EXHIBIT 3.10 Project Data Sheet

Project:

Address/directions from main office:

Site entry from the road/highway:

Project manager:	[Name]	Day: Night: Cell:
Project engineer:	[Name]	Day: Night: Cell:
Superintendent:	[Name]	Day: Night: Cell:
Owner/Developer:	[Name]	Day: Night:
Architect:	[Contact]	Day: Night:
Engineer:	[Contact]	Day: Night:

Consultants:

Major subs:

Hospitals/Paramedics:

Date of last injury involving lost time:

Date of last safety inspection:

(*Source:* Janine Reid Group, Inc.)

Foreign language expertise—mandatory

Because of shifting demographics in the United States and the emergence of global civilization, all industries must prepare themselves for multilingual workforces; this need will only increase. Should a non-English-speaking employee be injured or killed, there is a strong possibility that the employee's family cannot speak English. The time to prepare for that event is *now*. Compile a list of salaried employees who can speak different languages and are willing to be called on for translation in a crisis. This list should include the employee's name, the language spoken, and his or her office and home phone numbers.

Emergency procedures contact list—mandatory

As discussed earlier, calm heads may not prevail when a crisis strikes. Therefore, detailed instructions need to be posted at strategic locations in the corporate office as well as at job sites. Exhibit 3.11 illustrates an emergency procedures card to be posted at all job sites. Use this outline as a guide to develop your own emergency procedures card.

Incident report—mandatory

As time marches on, memories fade and details of the crisis become vague. Therefore, it is recommended that the team leader make certain that an incident report is completed and circulated to the crisis management team as well as retained in the project file. This report reflects the information gathered at the outset of the crisis. Exhibit 3.12 illustrates what an incident report might look like for a crane-related accident.

EXHIBIT 3.11 In the Event of an Emergency

In the event of an accident, every effort should be made to ensure maximum safety for everyone. Only authorized personnel should be admitted to the scene.

FOR ALL EMERGENCIES DIAL 911
FOR IMMEDIATE EMERGENCY RESPONSE

NEXT, REPORT THE EMERGENCY TO ONE OF THE FOLLOWING:

Team Leader   [name]

Day:
Cell:
Night:

Backup Team Leader   [name]

Day:
Cell:
Night:

Safety Manager   [name]

Day:
Cell:
Night:

(*Source:* Janine Reid Group, Inc.)

EXHIBIT 3.12 Sample Incident Report

M E M O

TO: Crisis management team members and the project file

FROM: [Team leader]

DATE: [Date]

RE: Incident report for the QC Hotel Project Accident on [Date]

Below is the sequence of events as told by Joe Hardhat, crane operator, for AA Crane Company on [Date] at 10:25 A.M.

The crane had just completed a lift and was inactive. Joe heard a noise in the rear of the crane that sounded like a bearing. He had heard the same noise earlier in the day. Now that the crane was inactive, he took a minute to go back and check on the noise. He had positioned the crane boom in a place where it was up and out of the way.

Joe suspects that, as he got up to check on the noise, he hit the boom lever and accidentally activated the boom. He was in the rear of the cab for about one and one-half to two minutes. He turned around to go back to his seat and saw his radio power pack explode. He then looked out the window to see the project superintendent waving his arms and pointing at the boom. At that point, Joe realized what had happened. The boom had dropped slowly and the load line came in contact with the high-voltage line over a major intersection. Upon contact, the area was illuminated and there was an explosion and a small fire broke out. The electrical wiring for the temporary lighting was blown out. The headache ball was lying in the intersection and the project superintendent was trying to block traffic until the problem could be corrected.

Joe immediately put the boom lever in the neutral position and then began to return it. In doing so, he hit the high-voltage line again and caused a second explosion. Joe stopped for a minute and then continued to lift and succeeded in getting the cable out of the lines and cleared of the area.

There were no injuries and the small fire that resulted from the explosions was extinguished immediately.

Channel 9, who offices are less than two blocks from the site, had a film crew at the intersection within minutes of the incident. They did not try to enter the site nor did they attempt to interview anyone. They filmed the intersection with the headache ball as well as the job site. A report of a brief power outage, plus film of the intersection, was aired on the 6:00 P.M. and 10:00 P.M. news but there was no mention of our name or the project owner's name.

(*Source:* Janine Reid Group, Inc.)

Evacuation plan—mandatory

A clearly defined plan for evacuating the corporate office should be included in your crisis management plan as well as posted in a highly visible area. Employees must be briefed on evacuation protocol and their respective responsibilities.

What should be included in a project plan?

Two options are offered in this section. The first option concerns very large projects and projects distant from the home office that must be able to handle crises on their own without physical support from corporate headquarters. For those projects, the senior person on site should retain a complete copy of a crisis management plan. At a minimum, the following components should be included:

Cover sheet
Control sheet
Table of contents
First-hour response checklist (customized to the project)
Contacts in the event of an emergency (customized to the project)
Notification procedures in the event of a serious employee injury
Notification procedures in the event of an employee fatality
Procedures for a fire, bomb threat, or hazardous materials spill
Tips for working with the news media
Project data sheet (customized for the project)
Foreign language expertise (customized for the project)
Emergency card (customized for the project)
Evacuation plan (customized for the project)

What should be included in a project plan for short-duration projects?

Some companies may feel as though they do not need a plan for short-duration projects, but remember that a crisis can occur anytime, anywhere, to any size project. It is not necessary to have a complete project crisis management plan, but the senior person on site should have the following components:

First-hour response checklist (customized to the project)
Contacts in the event of an emergency (customized to the project)
Notification procedures in the event of a serious employee injury
Notification procedures in the event of a fatality
Procedures for a fire, bomb threat, or hazardous material spill
Tips for working with the news media
Foreign language expertise (customized to the project)
Emergency card

How should your plan be organized?

Typically, your plan is accessed in a time of urgency. Therefore, the most vital components should appear in the first few sections of your plan. These may include your first-hour response checklist and contacts in the event of an emergency. The remainder of the plan should be organized in a fashion that makes logical sense to your company and crisis management team.

Your plan is a living document; additions, deletions, and changes will occur periodically. A three-ring binder accommodates these changes. The front and the spine of the binder should clearly indicate its contents, and each section should be have its contents identified on a tab for quick reference.

Who should get a copy?

The core crisis management team members, as defined in Exhibit 2.1, and their backups, should each receive two copies of the corporate crisis management plan. Each should keep one copy at the office and the other at home because crises can occur during the day, night, or weekends. In addition, the safety manager, human resources, and the team administrator should also be assigned two copies of the plan—again, one for the office, one for the home.

The senior person at each of your projects should also be assigned office and home copies of the project crisis management plan, customized to the project. If mobile a large percentage of the time, this individual may wish to photocopy selected components of the plan to keep in his or her work vehicle.

What type of event calls for the plan's activation?

The team leader should activate the crisis management plan at the first sign of any event that has the potential of disrupting normal business activities or generating attention from outside groups. Better to err on the side of caution than to do nothing and then have to play catch-up in a crisis.

How often should your plan be updated?

There is nothing more frustrating than needing a contact name or phone number in crisis and then finding out that the number has changed or is no longer in service. Thus, as a living document, your crisis management plan must experience changes. The team administrator should review, update, and distribute new information to the crisis management team quarterly. The last revision date should be printed on the bottom of each page in your crisis management plan.

Can a crisis management plan handle all of your crises? Of course not, but a plan helps you make better decisions under pressure because you have thought through the process prior to the event. Luck truly does favor the prepared!

Chapter Summary

- The first-hour checklist is a critical component in your crisis management plan because it assists the team members in gathering facts and identifying steps that need to be taken at once.

- A variety of resources must be contacted and mobilized when a crisis strikes. Make certain that your list of contacts is comprehensive and that the phone numbers are current. It is useful to put the key contacts on a laminated sheet the size of a business card; this can then be carried by each crisis management team member.
- It is recommended that a crisis command center be established either on or off site. The center will serve as a central meeting area for the crisis management team as well as authorities providing assistance.
- All of the pages in your crisis management plan should indicate the date of the last revision. The team administrator is responsible for reviewing and updating the plan quarterly.
- Rapid notification of the spouse/family of a seriously injured employee or of a fatality should be a top priority. There are no excuses for any type of delay in this area.
- Company history and project information should be written prior to a crisis so they may be presented in a thoughtful, positive manner.
- Know your skeletons. Be prepared to answer questions relative to past incidents that may put your company in a negative light.

4

HOW REPORTERS DO THEIR JOBS

Where do they get those headlines?

OAKLAND CAVE-IN KILLS WORKER: FELLOW LABORER DECAPI-TATES BODY IN RESCUE ATTEMPT (*Detroit News,* 1999)

WORKER IS KILLED IN COLLAPSE LINKED TO ILL-RIGGED SCAF-FOLD (*New York Times,* 1996)

FATAL CAVE-IN AT WORK SITE RAISES QUESTIONS ABOUT SAFETY; FAMILY SAYS WORKER HAD MISGIVINGS BEFORE TRAGEDY IN ANTONIA (*St. Louis Post-Dispatch,* 1999)

THIRD FATALITY STRIKES LAS VEGAS CASINO (*ENR,* 1999)

UTILITY CREW PUNCTURES GAS LINE, EXPLOSION KILLS 4 (*ENR,* 1998)

CONSTRUCTION WORKER BURIED ALIVE IN PIT (*New York Times,* 1998)

ACID INJURES WORKERS AT REFINERY WHERE 6 DIED (*Seattle Times,* 1999)

GROSSLY INCOMPETENT DESIGN CENTRAL TO WALKWAY DISAS-TER (*Safety & Health Practitioner,* 1999)

Headlines like these can be a company's worst nightmare. If something is leaking, flaming, exploding, hurting, dying, obstructing, delaying, decaying, polluting, or falling, it will likely draw the attention of the news media. The result could trigger a barrage of nasty images and headlines.

The purpose of the headline, or *news tease,* in a broadcast, is to draw readers/viewers into the story. Obviously, a headline or news tease that screams death and destruction will garner the desired interest. Unfortunately, an entire story cannot be told in the headline, which can, therefore, create misunderstandings by those who choose not to read the text or listen to the story. The implication of the headline can thus deal an unfair blow to the company in crisis.

The reporter covering your crisis does not write the headline. That task is assigned to a copy editor, who bases the headline on two things: (1) a quick read of the story and (2) the amount of space/time allotted the story. According to Lynn Bartels, a reporter for the *Rocky Mountain News,*

> copy editors write the headlines and sometimes reporters go berserk when they see the headlines. Copy editors do not deliberately write inaccurate or inflammatory headlines, but unfortunate headlines do appear. Should this occur, I would recommend that a company representative contact the reporter and acknowledge the fact that the reporter did not write the headline and inquire who should be contacted to lodge a complaint. (Personal communication, April 16, 1999).

Because you have no control in this area, let's move on and discuss how reporters do their jobs and how you can exert some influence over their coverage.

Why is your worst day a reporter's best day?

> A journalist is someone who cannot distinguish between a bicycle accident and the end of civilization.
>
> —George Bernard Shaw (circa 1890)

I often wonder how Shaw would characterize the media today!

Many companies and consultants feel that a crisis is an easy target for reporters who typically require little initial research to cover the crisis and usually provide provocative images to generate reader/viewer interest. A media outlet hears about a problem, dispatches a crew to the site, points a camera at the problem, and asks questions of people who are in a highly charged emotional state. Then the story tells itself. According to James E. Lukaszewski (1994), chairman of The Lukaszewski Group, Inc., "Today's journalism is a combination of speculation, sensationalism, fragmentary information, dramatization and the dissemination of unverified information" (p. 37).

Has the line been blurred between objective reporting and sensationalism? There must be another side to this story.

Patrick O'Driscoll, *USA Today's* Denver bureau chief, has this to say on the subject of companies in crisis:

> Of course, a company in crisis makes for a good story—but not because news reporters and editors delight in a company's misfortune. A crisis, by nature, involves drama, emotion, conflict and other elements that usually turn an event into news. That doesn't mean a company in crisis automatically will make news. But think about it: In a major con-

struction accident, employees are affected, customers are affected, stockholders are affected, management is affected. And often, the rest of us are affected. (Personal communication, April 16, 1999)

When your crisis strikes, a reporter's goal is to find out what happened and who is responsible. Many companies feel as though journalists focus too much energy on the negative side of a company in crisis. Is this focus the result of reporters viewing themselves as the public's watchdogs and, therefore, thinking their responsibility is to uncover newsworthy stories? Is it due to their continual quest for higher ratings in an extremely competitive industry? Is it that corporate America provides the media with ample fodder to report on these stories? No matter the reason, reporters are not going to go away if they feel there is a story to be told. And each wants to tell that story before anyone else does. With this in mind, let's discuss how a company in crisis can assist a reporter and, at the same time, protect its best interests.

What are reporters looking for when they cover your crisis?

Information and answers—period. What do you lack most during the outset of a crisis? Information and answers. Here lies the problem.

Communication with the outside world during the first few hours of a crisis is both vitally important and extremely difficult. Reporters want to know what happened, who is involved, who is to blame, and what you are doing about it—and they want to know NOW.

In my interview with Patrick O'Driscoll of *USA Today,* I asked what reporters are looking for when they cover a crisis. His informative response:

Answers. Many times, though not always, company-in-crisis stories are breaking-news events on a tight deadline which surely would apply to a major construction accident. We reporters need basic information quickly. In one more word, we are also looking for *access.* Reporters need to reach the people who can best answer questions as soon as possible, without a lot of background noise. We need to visit the scene—as long as we are not endangering ourselves or getting in the way. A common reaction is to shut down the site and ban everyone. You would be surprised how much more cooperation you would get if you led us in, in small groups or even on a pool basis, to see for ourselves, and for our stories, where it happened.

There's not a lot of time to spend going over the finer side points of every last detail about the company or the issue. Get to the point: what happened, safety systems in place, equipment issues, condition of injured workers, effect on the construction schedule or, if a construction site is in the middle of a city or neighborhood, what effect the surrounding neighbors may need to worry about. (Personal communication, April 16, 1999)

O'Driscoll makes a point worthy of further comment. Once the initial rush of the crisis is over, typically after the first one to two hours, you must be prepared to address the key points of interest. This necessity raises the questions "How do I identify key points, and when can/should I release that information?" Put yourself in the reader's/ viewer's seat and determine what information would be important to you. After you have made that determination, list the items that you can confirm and get approval for release to the news media. This information will give you the basis for your statements. By no means should you disseminate any information that has not been confirmed and approved by upper management for release. This topic is discussed in great detail in Chapter 5.

I wanted to see if the broadcast media felt the same as the print media about the key points they wanted covered in a crisis, so I interviewed Tracy Berry, who has over 20 years of radio and television reporting experience. She is currently the executive producer and director of program development for Chambers Productions in Eugene, Oregon. She indicated that the key points in a crisis are extensive:

> I am looking for basic, accurate information, such as: What happened? How did it happen? Who is affected? Is there any continuing risk? What are you doing to take care of this problem and prevent its continuance/reoccurrence? If the crisis is still in progress, how much longer will it last and what will it take to resolve it?

> As time allows, I will try to put the crisis in perspective by asking a whole host of questions: Has this happened before with this company, or this industry, or in this community? How does it affect the people in the firm, or the industry, or the community? I will try to track down people directly affected by the crisis to get their perspectives on what occurred and what can be done to prevent a future occurrence. I will take a look at economic impacts and try to determine whether there are any lessons to learn from what happened. Can other companies use what occurred to review their own policies, or guidelines, or safety records, etc.? How is the community responding to this crisis? Does this change the way we view our community? Does this shape debates over various public policies and laws? These are but a few of the questions that initially come to mind. (Personal communication, April 29, 1999)

What can a company in crisis do to help the reporter cover the story yet protect its best interests?

"It depends on the story," explains Lynn Bartels of the *Rocky Mountain News* in my interview with her. For companies in crisis, she offers the following advice:

> In general terms, have fact sheets available about your company, see if co-workers would be willing to talk to the press and be photographed or videotaped, have someone with a pager who responds to calls, be mindful of deadlines, and so on. (Personal communication, April 16, 1999)

John Mandes, assistant professor of mass communications and journalism studies at the University of Denver and past reporter for the *Albuquerque Tribune* and the *York Dispatch,* has a succinct answer to this question: "Tell the truth! Never say, 'No Comment'—never, ever!" (Personal communication, April 16, 1999).

Broadcast journalist Tracy Berry adds,

I cannot stress strongly enough the need for *accurate* information in the early hours of the crisis. Do not speculate. Please do not make things up. Do not try to misdirect us with intentional misinformation. Be aware that there are other sources of information out there. Conflicting reports create confusion and, if the crisis is still underway, can create additional problems. If you cannot comment, tell me why and then tell me when you can comment. A fact sheet on your company is always helpful. (Personal communication, April 29, 1999)

Patrick O'Driscoll of *USA Today* offers additional insight:

A common reaction by some companies to protect their best interests is to stonewall, which creates the single biggest issue separating news people and those they cover in a big crisis. Or to delay, delay, delay. Or to address anything but the issue at hand. My advice, invariably, is to come clean if there is a problem and to be open and candid even if there isn't. A company that is doing something to fix the problem always comes across better than one that avoids, evades, obfuscates and downplays. Yes, in these litigious times (especially surrounding a major accident involving bodily injury/death, contract liability, and the like), there are plenty of lawyer-initiated reasons not to say a damned thing, but trust me, it only makes things worse to clam up completely. (Personal communication, April 16, 1999)

Just as no two crises are the same, neither are any two reporters. Your crisis may occur in a remote location where only one reporter from the local paper arrives and delivers a fair and balanced report on the event. On the other hand, your crisis may occur in a major metropolitan area with national and international media representatives who will demand answers to difficult questions within minutes of the outset of the crisis.

Always be prepared for media interaction in a crisis situation and do not try to second-guess on their view of a newsworthy story. Many reporters claim that "news" is whatever they say it is! As evidenced in everyday reporting, "importance" does not necessarily mean that a story is newsworthy; on the other hand, something that has an "interesting" slant could generate provocative news coverage.

Here are some do's and don'ts of working with the news media in a time of crisis.

- *Do talk.* Saying little is better than saying nothing. Explaining why you cannot talk is better than stonewalling. If you want your side of the story told, you must tell it. If you do not tell it, reporters will get a version elsewhere—perhaps from a disgruntled employee who was laid off the week before or a worker who has just witnessed his or her best friend getting hurt or killed.

- *Do tell the truth.* Reporters will find it out anyway, so be honest and accurate when you are giving information. This advice does not mean you have to give every detail, but be truthful. If you do not know the answer, say so! It's not a crime to say "I don't know" or "I'm not absolutely certain about that," as long as you follow such statements with "but I'll find out and get right back with you." *Do respond quickly.* If you do not, the wrong story may be told and that is tough to erase.

- *Do emphasize the positive and communicate your corporate message.* Remember to emphasize the good safety measures taken, the damage limitation due to good teamwork by your employees, and the steps the company is taking to minimize the effect of the emergency on the community.

- *Do stay away from liability issues.* Do not talk about who is responsible, do not make accusations, and do not give out company or individual names. Whatever you say may become part of a legal case, so be as general as possible.

- *Do take control.* If there is bad news, release it yourself before a reporter digs it up and tells the world.

- *Do create visual analogies.* The old saw "a picture is worth 1,000 words" applies here. Give examples, such as "The affected area covers 40,000 square feet, which is the approximate size of a football field."

- *Do condense your information.* The average sound bite is just seven to ten seconds long, so try to keep your response to no more than two to three sentences. The first sentence should be your direct response and the next one or two sentences should support or explain that response.

- *Do make sure your information is accurate.* It should come from a reliable source and you should understand the details thoroughly.

- *Do make sure reporters know who the spokesperson is.* The corporate spokesperson should be the only person authorized to disseminate information to the outside world. It is important that the company "speak with one voice." Keep in mind that no information should be released without approval by upper management and legal counsel.

- *Don't say, "No comment."* This statement implies guilt. If you do not know the answer to a question, tell the reporter you do not know but will try to find out. If the question may lead to an embarrassing answer, give as much information as you can in as positive light as possible. If you make a mistake, admit it. Avoid excuses. Explain how you're planning to make things right.

- *Don't be trapped into predicting the future and* never *speculate!*

- *Don't say anything off the record.* If you do not want a statement used, do not say it.

- *Don't wear sunglasses when being interviewed.* You will be perceived as shifty and as hiding something.

- *Don't discuss damages or estimated costs.*

- *Don't discuss facts relating to insurance,* such as amounts and terms of coverage, name of carrier, possibilities of settlements, or reimbursements.

- *Do think before answering.* Taking time before you respond is perfectly acceptable. You, not the reporters, are in control of your responses. Do not allow them to rush you. If you do not understand questions, ask the reporters to rephrase them.

Communicating in the first 24 to 48 hours of a crisis is critical and information rarely plentiful. This scarcity does not give you permission, however, to avoid the news media. If there is a story to be told, reporters will tell it—with or without you. Here is an exercise you should try: Watch the news coverage of a crisis in progress and critique the response of the affected company. Decide whether you agree with its approach or would handle matters differently. Evaluate the company's spokesperson for response time, clarity of message, position strategy, apparent honesty, and consistency. Then talk to your colleagues and get their feedback. This exercise affords you the luxury of learning from someone else and, quite honestly, there are plenty of teachers out there serving themselves up daily for your educational benefit.

Take control of your destiny and learn how to work with the media because they have the ability to control how your audiences perceive your actions during a crisis. You must convince the media that your side of the story is valid and that you are a reliable source of information. Your job is to make certain that the best possible outcome occurs.

Chapter Summary

- The purpose of a headline is to create interest in the reporter's story; however, neither you nor the reporter has any control over the headline's wording. If you feel a headline for your story is inaccurate, contact the copy editor and lodge a complaint.

- The definition of "newsworthy" can be controversial. Remember that if reporters feel your crisis is worth covering, they will tell the story—with or without you. Therefore, be proactive from the outset and communicate your side of the story.

- Reporters are interested in answers to the proverbial who, what, when, how, and why questions. If you cannot confirm information for release, simply tell them why and advise them when the information will be available.

- Never speculate or say "No comment." Absolutely no information should be released until it is confirmed and approved for release by upper management.

- Information on the company should be at the ready for distribution to the news media. This includes a fact sheet on the company and the project (if the crisis is project-related), information on the company's safety history, and other background information you feel may be relevant.

- Crises occur throughout the world every day. Watch the media's coverage of them closely and learn from the wide variety of responses by the companies in crisis. Firsthand experience is obviously the best teacher, but learning from others is far less stressful.

5

WORKING WITH REPORTERS IN A CRISIS SITUATION

Chapter 4 provided background on how reporters do their jobs and what they are looking for when they cover your crisis. We can now proceed to the actual process of working with reporters to communicate your message(s) during a crisis.

A hypothetical construction company in the midst of a crisis has been created to illustrate this chapter.

Background

XYZ Construction Company is a privately held general building contractor headquartered in Denver, Colorado; it also has district offices in Southern California and Arizona.

XYZ Construction Company has a strong safety program, which is enforced by a full-time safety director. Last year, however, the company experienced an accident on a school project in Phoenix that resulted in two serious injuries. OSHA levied a heavy fine against the mechanical subcontractor responsible for the incident.

The Project

XYZ is currently at work on a ten-story medical office building located just 20 minutes from its corporate office in Denver. This big project offers a strong possibility for additional business with the building's owner if all goes well.

The Neighborhood

The project is located near a residential area. Ever since the survey crew set foot on this project site, problems have arisen with the residents—major problems. In fact, they elected a group leader who has been relentless in trying to stop the project.

The residents feel that their neighborhood is not the right location for a medical office building and that alternative sites were not properly investigated. Every day, complaints are called in to the site about the noise, dust, and truck traffic. XYZ has tried to be a good corporate citizen. Hours of operation have been changed and air measurements and daily water trucks to hold down the dust have become the norm. The residents were satisfied for a while, but the group leader continues to work against the project.

Some Problems

Everything has been running smoothly, except for two things. First, this month Denver received a record rainfall, which resulted in a schedule delay of two weeks. Some workers were overheard commenting that the superintendent has been compromising their safety in an effort to catch up. Second, Mary Smith, who works across the street from the project, has contacted the corporate office on several occasions complaining about the behavior of the construction workers. She claims that the workers make lewd gestures and remarks as she walks by the project to work each day.

Today—at the Project

On Tuesday afternoon, the mechanical subcontractor used the tower crane to lift a cooling tower to the roof of the building. As the tower approached the eighth floor, a loud crack was heard; the cooling tower broke loose from the rigging and crashed into the ground, hitting two of the mechanical subcontractor's employees as well as an XYZ employee.

A brief silence fell across the job site, and then the screaming started. Someone called 911 and within five minutes EMS and the police arrived. A reporter was close behind.

The XYZ employee is presumed dead and the other two workers are seriously injured. The job site is in total chaos. A reporter is trying to enter the job site trailer to get an interview.

Let's take a moment to analyze this scenario. XYZ is the general contractor on this project and is the highest entity in charge (Exhibit 2.2); it is, therefore, responsible for implementing its crisis management plan. As you can imagine, XYZ's actions during the first few hours of this crisis will be scrutinized by its audiences and the owner of the project. Such scrutiny is difficult because, during this time rumors, speculation, and emotion are at their peak. XYZ, however, has a crisis management plan and begins to mobilize its crisis management team. The first priority of the team is to secure the site and make certain that all employees are accounted for. Next, the team will address the action items listed in Exhibit 3.1.

Right now, the team is feeling overwhelmed with all of the demands that the crisis is placing on them, and the last thing they want to confront is a pushy reporter. A crisis creates a tremendous amount of pressure, so the crisis management team may want to delay the process of speaking to the reporter as long as possible. Such a delay may not be the correct decision.

Clearly, a crisis generates chaos, which, in turn, creates an urge to stonewall the media. After all, XYZ has a site to secure and people to take care of, so the natural response is to say "No comment." However, this response screams "Guilty as charged!"—Whether or not a charge is specified. Another thought that enters the minds of the crisis management team is, "To hell with reporters; they can wait!" Of course they can wait, but are a company's best interests served by this attitude? Case study after case study shows that reserving comment during the critical first hours of a crisis does not present a good company image in the media. Another option that the crisis management team may explore is issuing a statement that comments will be withheld until all of the facts of the incident are discovered and verified. This course of action could present a problem from a communication standpoint because the investigation process in a crisis of this size could take considerable time.

Finally, some companies think that if their projects are located in remote areas, reporters will not find out about accidents. If, by chance, a reporter does find out about an incident, a company might mistakenly believe that the coverage will be restricted to the local media. Unfortunately, that line of thinking is not an option in today's technologically endowed world. According to John Scanlon (1999), crisis manager for DSFX, "The rolling news day has enhanced drive-by journalism. The change has been so dramatic that distance is dead, place is moot and time has disappeared. Fiction travels as fast as fact, and there is no such thing as a local event" (p. 18).

In our world of 24-hour news and instant Internet access, our goal in crisis communication is to be as open as possible as quickly as possible. To do otherwise raises the red flag that information is being hidden. This action could also cause reporters to become aggressive in their quest to discover that suspected hidden information. Understand, however, that speed should not sacrifice organization and preparation, which are our focus in the remainder of this chapter.

Perception is reality, and the first message of a company in crisis to its audiences is the most crucial. A company must project the image of proactivity and of control over the flow of verifiable information. But how do you do that when you have precious little information? You deliver a buy-time statement.

What does it mean to buy time with the media?

The buy-time statement is the one that the temporary spokesperson (as defined in Chapter 3) delivers as the first statement about the incident to the media. This statement accomplishes the following goals:

- Acknowledge the incident.
- Disseminate information that has been verified by a reliable source or sources.
- Buy time to gather more information and to get the corporate spokesperson on site to take over the responsibility of working with the reporter(s).

The team leader and corporate spokesperson assist the temporary spokesperson in the development of the buy-time statement. This statement, as well as all subsequent statements, is built upon a foundation comprising three major components:

1. Identification of verifiable information from a reliable source or sources and obtaining approval from upper management for its release.
2. Organization of the information and development of the statement based on a beginning, key points, and an end.
3. Anticipation of questions and development of responses.

These components form the basis of all statements; they are referred to frequently as we progress through our hypothetical crisis.

Now, it is time to return to our scenario and begin to build a buy-time statement. We start with our first foundation component, which is the identification of verifiable information from reliable sources and obtaining approval from upper management for its release.

As you remember, the incident has just occurred and a reporter is standing at the job site trailer trying to get an interview. You are the temporary spokesperson and need to develop the buy-time statement based on the following information, which has been verified by two sources:

- The accident occurred at 3:30 P.M. today during the lift of a cooling tower to the roof of the building.
- Three workers were injured and EMS is transporting them to University Hospital. The severity of the injuries is not known at this time and the notification of the families is in progress.
- All of the other employees are accounted for.
- The site has been secured. Authorities are on site and the investigation is underway.
- John Smith, the corporate spokesperson, is en route to the site.

The second foundation component requires us to organize the information and develop the statement based on a beginning, key points, and an end. Any presentation, whether written or oral, should feature a logical progression of information. Let's work through this second component one step at a time.

Organizing the information

To organize the material in a logical format, we use the tried-and-true what, when, where, who, why, and how questions.

What: A cooling tower broke loose from the tower crane during a lift to the top of the building.
When: At 3:30 P.M. today.

Where: The eighth floor.
Who: Three workers are injured and are being transported to University
            Hospital. The families of the injured workers are being notified.
How/Why: The cause is unknown at this time. An investigation is underway.

You are now ready to write your statement.

The beginning

You should always introduce yourself at the beginning of your statement. If you have a name that is difficult to spell, make certain that you spell your name for the reporter(s) because media outlets copy one another, if your name gets spelled incorrectly in one place, it could appear that way throughout the coverage of your crisis. Therefore, make sure your name is spelled correctly from the outset.

Next, always open your buy-time statement with the words, "Here is what I can confirm at this time." These nine words put the reporter(s) on notice that you have verifiable information to release in a statement format. Most reporters will then allow you to finish your statement without interruption.

Key points

Just as a building needs a supporting structure, so a solid statement needs to be built upon three to four key points, which typically address the cause of the incident, concern for the injured and their families, and the status of the job site. The key points of any crisis communications statement should address the primary interests of the specific audience you are communicating with. In this chapter, the audience is the news media; however, as we discussed in Chapter 2, you have a variety of audiences, so your message may vary slightly by audience. Chapter 8 discusses techniques for communicating effectively with your other audiences.

The following is an example of three key points to include in the buy-time statement for XYZ's crisis:

1. *Cause:* At this time, we do not know the cause of the failure; however, we are working very closely with the authorities to find out exactly what happened.
2. *Concern:* We are sorry to report that three employees were injured and are being transported to University Hospital. We are notifying the families and staying in close contact with the hospital.
3. *Status:* The site has been secured and all employees are accounted for. John Smith, our corporate spokesperson, is en route to the site. We will have an information update in one hour at (location).

The end

The end of your statement reviews what you are doing about the situation and the time and location of the next update.

The third foundation component requires that we anticipate questions and develop responses. This component is described in detail in the next section.

Now, based on the information that has been gathered and verified, as well as our identification of the beginning, key points, and end, our statement might read as follows:

> My name is Justin Reed—that's J-u-s-t-i-n R-e-e-d—and I am the superintendent on this project for XYZ Construction Company. Here is what I can confirm at this time.
>
> At 3:30 this afternoon, while being lifted by the crane to the roof of the building, a cooling tower broke loose at the eighth floor. We do not know the cause of the failure and are sorry to report that the accident resulted in three employees being injured. Those employees are currently being transported to University Hospital, and our immediate focus is on notifying the families and determining the extent of the injuries. We will release the names of the injured workers as soon as the notification process is completed.
>
> Again, we do not know the cause of the failure, and we are working closely with the authorities to find out exactly what happened. The site has been secured and all employees are accounted for. John Smith is our corporate spokesperson, and he should be arriving shortly. We will both be your points of contact. If you elect to stay on site until the next update, we ask that you remain in this safety area. Right now, I need to return to the site to provide assistance and gather more information. John Smith or I will return in one hour to provide you with updated and verified information.

In analyzing this statement, note that the content is sketchy and merely verifies what the reporter can see. However, the statement nevertheless accomplishes two things:

1. It establishes a working relationship with the reporter(s) and shows that you will be forthcoming with new information as it becomes available and is verified.
2. It allows you to buy one hour of time to gather additional information for the next interview.

One of the most critical parts of the buy-time statement is the end. When you have finished your statement, walk away and *do not* entertain any questions. Remember that the crisis has just happened and verifiable information is almost nonexistent. You, as the temporary spokesperson, are vulnerable to potentially explosive questions that cannot be answered at this time, such as:

How could you let this happen?
Who is responsible for the accident?
What is your company wide safety record?
Can we interview the witnesses/victims?
What are the names of the injured/killed?
What is the potential for continued damage or danger?
What is the estimated damage in dollars?

Will you shut the project down and, if so, for how long?

Can we get in to take pictures?

Walking away may feel very uncomfortable; indeed, for many a spokesperson, it is analogous to hanging up on someone over the telephone. If this gesture proves too uncomfortable, as reporters yell out a series of questions, simply say, "That's all I can confirm at this time. Thank you for your patience, and either John Smith or I will be back in one hour." *Then* walk away.

Questions must be addressed at future media interviews, but this territory is covered in the next section. Remember, questions should not be entertained after the buy-time statement. Period.

If you are the spokesperson for your company, whether the temporary or the corporate spokesperson, you will feel nervous when you deliver your statement to a reporter. This is normal and should be expected. Even the savviest spokespeople get the jitters and risk the possibility of getting off track, blanking out, or simply rambling when they should be confident and succinct. Therefore, you should *read* the statement to avoid such problems. This technique also helps to relieve some of the pressure caused by anxiety. The statement can be copied and distributed to the reporter(s) to minimize the possibility of misunderstandings or misinterpretations that can result from an oral presentation. Background information on your company and project can also be included (as explained in Chapter 3).

How do you develop statements as the crisis progresses?

We are nearing the one-hour point of our hypothetical crisis, when the temporary spokesperson has promised the reporter(s) updated information. Now it is time to practice what we have learned.

As a crisis progresses, additional information becomes available, and the corporate spokesperson must be able to communicate that information quickly and succinctly. Delivering the information quickly is a difficult job in and of itself, but brevity is one of the toughest things that can be asked of a spokesperson in the construction industry because it flies in the face of years of training.

Construction is a highly technical and detailed industry, and many industry spokespersons reflect the world in which they work when they deliver a statement to reporters. Their statements become wordy in an effort to describe the event in detail—sometimes excruciating detail. This tendency can, and does, create a problem with the media, because reporters have a limited amount of space/time to tell the story. If a spokesperson uses 800 words to describe an issue but the reporter has space for only 150 words, something has got to go. Would you trust a reporter to reduce your 800 words to 150 and still maintain the integrity of the original content? Probably not. Yet such compression is exactly what happens, and then the company spokesperson screams, "I was taken out of context" or "I was misquoted." Many times, the spokesperson goes into too much detail trying to educate rather than communicate, and the reporter is forced to reduce the information to accommodate the space/time allotted to the story.

A crisis is not the time to educate a reporter on the technicalities of your company and the construction industry. There simply is not time or space. Learn, and practice, reductionism by structuring each of your key points not to exceed two to three succinct sentences and put the most important fact in the first sentence. To be blunt, get to the point quickly. To do otherwise is begging for misinterpretation. Remember, if your local television station can report the world news in less than 15 minutes, you can tell your story in three to four key points.

Let's continue with our crisis at the XYZ job site. John Smith, XYZ's corporate spokesperson, has arrived at the job site and is putting the finishing touches on his statement. Review the process John took in developing this statement.

It is 4:15 P.M. and an information void continues to exist; however, bits and pieces of information have been discovered and upper management has approved the release of this information to the media. John Smith begins building his statement by identifying verifiable information. For the sake of illustration, assume that the following information has been verified and approved by upper management for release:

1. The families of the injured workers have been notified and are being transported to hospital. At this point, names can be released.
2. Juan Lopez, a laborer for ABC Mechanical Company, has a broken leg and will be released from the hospital this evening. Ron King, also a laborer for ABC, has internal injuries and is in serious but stable condition. Larry Lynn, a carpenter apprentice for XYZ, suffered major head injuries and is in critical condition.
3. The site will be shut down until further notice.

The crisis is only one hour old, so the traditional what, why, where, who, when, and how questions remain the same as described in the buy-time statement.

John Smith needs to develop the key points that address cause, concern, and the current status of the incident. Again, the crisis is only one hour old, so much of the information that was used in the buy-time statement remains the same.

1. *Cause:* We do not know the cause of the failure, but we continue to work very closely with the authorities and will not rest until a determination has been made.
2. *Concern:* The families of the three injured workers have been notified and are en route to University Hospital. We will remain in constant contact with the hospital and the injured workers' families throughout their recovery.
3. *Status:* The job site will be closed until further notice. We will have an update at 9:00 A.M. tomorrow morning at [location].

The end of each statement should give the status of the incident and the time and location of the next update.

John Smith's statement reads as follows:

My name is John Smith and I am the [title] for XYZ Construction Company. Here is what I can confirm at this time. At 3:30 this afternoon we experienced an accident that involved a crane lifting a cooling tower to the roof of the building. A failure of unknown origin occurred that caused the release of the cooling tower at the eighth floor. This is a tragic accident, and we are sorry to report that three workers were injured as a result of this accident. We have contacted the families of the three workers who were injured, and they are currently at University Hospital.

I have the names and latest condition of the injured workers. Juan Lopez, who is a laborer for ABC Mechanical Company, has a broken leg; we are happy to report that he should be released this evening. Ron King, also a laborer for ABC, has internal injuries and is in serious but stable condition. Larry Lynn, a carpenter apprentice for XYZ Construction Company, suffered head injuries and is in critical condition. There are no words to express our sadness regarding these individuals, and our focus will remain with them, and their families, through the recovery process.

As I mentioned, the cause of the accident is unknown at this time. We are cooperating fully with the authorities to determine exactly what happened so we can make certain that it never happens again.

The job site will be shut down until further notice. That is all of the information that I can confirm at this time. I'm sure you can understand that we need to deal with the emergency at hand and gather more verifiable information. I will have an update at [location] for you at 9:00 A.M. tomorrow morning, and I will then be prepared to answer your questions.

John Smith should walk away after this statement. Clearly, reporters will not appreciate the fact that John will not entertain their questions; however, a question and answer session at this juncture could be uncomfortable because of the lack of information. John Smith will have his hands full getting ready for the interview at 9:00 A.M. the following day, when he *must* be prepared to answer questions. Let's fast-forward and assist John in the preparation of his next statement and get him ready for the question and answer session that will follow.

Beginning with the first foundation component of building a statement, we assume that the following information has been verified and approved for release to the news media:

1. Juan Lopez, a laborer for ABC Mechanical Company, was released from the hospital at 6:00 last night with a broken leg. He is expected to make a full recovery. Ron King, also a laborer for ABC, has been upgraded to fair condition and has two broken ribs. Larry Lynn, a carpenter apprentice with XYZ Construction Company, remains in critical condition with a skull fracture.

2. The cause of the accident is still unknown. OSHA (Occupational Health and Safety Administration) is on site conducting its investigation, and XYZ Construction Company has hired a third-party investigation team to conduct an independent investigation.

3. The site has been secured and there is no potential for additional damage. The site will reopen tomorrow morning at 6:00.

We take a brief time-out here to talk about an important issue. You have probably noticed that one major thread is woven through the fabric of the first two statements and must continue with every statement when human life is involved. That is the thread of concern and compassion for those affected by the crisis. This concern must be the primary key point in all statements. Any company that avoids expressing concern and compassion for people hurt, killed, or inconvenienced appears cold or, worse yet, calculating. This perception must be avoided at all costs.

That important issue considered, John's statement at 9:00 the next morning may sound like this:

> Good morning. My name is John Smith, and I am the [title] for XYZ Construction Company. I have some additional information that I can confirm relative to the accident that occurred on our job site yesterday afternoon.
>
> First and foremost, our concern for the employees who were injured as a result of the accident continues. Everyone at XYZ Construction Company has been touched by this tragedy. I am happy to report, however, that Juan Lopez, who is a laborer for ABC Mechanical Company, was released from the hospital last night. Juan has a broken leg and is expected to return to work within a few weeks. Ron King, who is also a laborer for ABC, has been upgraded to fair condition, with two broken ribs and a sprained ankle. He is expected to make a full recovery. Larry Lynn, a carpenter apprentice for XYZ Construction Company, remains in critical condition with a skull fracture. We are staying in very close contact with the workers, their doctors, and the families because our employee's speedy recovery is our primary concern.
>
> We are also very concerned about determining the cause of this accident. As you recall, the tower crane was lifting a cooling tower to the roof of the building when a failure of unknown origin occurred at the eighth floor. We are diligently working with the authorities to determine the cause of that failure and will leave no stone unturned in that process. As a matter of fact, we have called in an additional investigation team to make certain that all of the facts are known. The site has been secured, debris has been removed, and we will reopen the project for work at 6:00 tomorrow morning.
>
> This is all of the information I can confirm at this time. I will be providing an update at 3:00 this afternoon with any new information that is verified. Until then, I can be reached at the office at [phone] anytime between [hours]. Now, I have just a few minutes to answer any questions you may have.

Notice that John Smith is willing to release his office phone number to the media. This is a wise move because it shows that he is willing to field reporters' calls and share whatever information he can. John also realized that it would not take much effort on the part of the reporter(s) to discover his phone number, so he elected to be proactive.

You may have also noticed that John ended his interview by saying that he would take a few minutes to answer questions. Now it is time to add our third foundation component—anticipating questions and developing responses.

What questions can you expect, and how should they be answered?

This is the part of working with reporters that many spokespeople dread because they know the cry of the media is, "Lay blame and justify!" Clearly, the question and answer part of the interview process can be unsettling; however, opportunities are available for the prepared. Preparation means anticipation, preparation, and rehearsal. The crisis management team should help their spokesperson anticipate all of the possible questions, prepare a response for each, and, if time permits, role-play a question and answer session. The team must then address three important issues prior to the interview:

1. *Determine the communications goal for the interview.* A spokesperson who stands in front of reporters without a clue about what the company wants to communicate is analogous to someone standing in front of a firing squad without hope of reprieve. You simply watch your life pass in front of your eyes. Believe it or not, your spokesperson does control the flow of information from your company to the outside world via the media, but that control quickly vanishes if a clear goal is not established.

   In our scenario, the crisis management team has established two goals: to communicate concern for the injured employees and to communicate a dogged determination to discover the cause of the accident. These goals will be illustrated in the form of key points in John Smith's next statement. Because of the wide variety of questions that John will receive, he may need to bridge back to his key points to make certain that his goal is being communicated. The bridging technique will allow John to transition, or build a bridge, from the question asked by a reporter to the response that he wishes to deliver. Examples follow:

   - I cannot confirm that at this time, but what I *can* confirm is . . . [back to a key point].
   - What people need to know is that . . . [back to a key point].
   - In order to put that into perspective, we need to talk about . . . [back to a key point].
   - It is important not to forget that . . . [back to a key point].
   - The proper protocol is not to discuss the findings until after OSHA's press conference, but what I can tell you right now is . . . [back to a key point].

   Always look for opportunities to transition to your key point(s) and do not be afraid to repeat your message at every opportunity. Who knows—a reporter may actually get it right if you say it enough times!

2. *Develop a mantra for the spokesperson.* John may be presented with a question or questions that he either does not know the answer to or that he is not comfortable answering. At this point, John needs to have a rehearsed answer—

a mantra—to get him through these rough waters. For example, if a reporter asks a question that John does not know the answer to, the mantra would be, "I don't know, but I will find out and get back to you by (time)." *Never* be afraid to say, "I don't know," but *always* follow those words up with "I will find out and get back to you by (time)."

Here are additional mantras that John could use in support of XYZ's key points:

- The safety and well-being of our employees is our first priority. We will leave no stone unturned in determining the cause of this accident.
- We will remain in constant contact with the injured employees, their families, and the medical facility. Our employee's full recovery is our top priority.
- We are fully complying with the investigation because we want to find out what happened to ensure that it never happens again.

A mantra should be developed for every interview, its content based on the individual crisis and the anticipated questions.

3. *Anticipate the questions that may arise.* Anticipating all of the questions that could arise is difficult, but certainly the most obvious as well as the most difficult questions can be identified and responses can be developed and rehearsed. This procedure will help John deliver the information in a manner that reflects confidence and control—even though he may not be feeling that way inside.

As we return to our hypothetical interview at 9:00 the morning after XYZ's accident, the crisis management team and John Smith have anticipated the following questions that reporters might ask:

How could you let this happen?

Who is responsible for this accident?

Your company had an accident in Phoenix last year that was caused by the same mechanical contractor you have on this job. Didn't you learn anything from that accident?

I understand that this project is behind schedule. Do you think this accident was caused by pushing your workers too hard to make up for lost time?

We had a conversation with a lady who feels as though your workers have harassed her as she walks to work everyday. Could this accident have been caused by your workers paying more attention to the women walking by the site than to their jobs?

Will you step up safety measures as a result of this accident?

What is the estimated damage in dollars?

Can we get in to take pictures?

As the crisis management team develops a response to each of the anticipated questions, it is important to reinforce that XYZ cares, is concerned, and is doing something about the situation. In addition, John must remember to practice brevity. Answers should be restricted to two sentences: one sentence should answer the reporter's question; the other should explain or elaborate on the answer. Brevity is vital because these answers may be used as quotes in a newspaper or sound bites on television and the radio. The typical quote in a newspaper has fewer than 20 words, and the average sound bite on television and radio is less than eight seconds long. That is brevity!

With conciseness in mind, let's review the questions that the crisis management team and John have anticipated. We will then discuss the reporter's perceived intent with respect to each and conclude with a company response. Keep in mind that the accident is less than 24 hours old, so information will continue to be sketchy.

Question: How could you let this happen?

Discussion: Do not fall victim to this question because it is loaded and leads nowhere—except to big trouble; however, it is frequently asked, so you must be prepared for it. Keep your communication goal in mind and refer to the key point that addresses cause.

Possible Response: The cause of this accident is unknown at this time. We will leave no rock unturned in finding out exactly what happened.

Question: Who is responsible for the accident?

Discussion: This is another fishing expedition for a cause. Questions that are rephrased and repeated should be welcomed because you have a response at the ready. All you need to do is to repeat your key point, as illustrated above.

Possible Response: The cause of the accident is unknown at this time. A thorough investigation is currently underway, and we will be better able to answer that question when the investigation is completed.

Question: Your company had an accident in Phoenix last year that was caused by the same mechanical company you have on this job. Didn't you learn anything from that accident?

Discussion: Another trap has been set for you. In Chapter 2, we talked about reporters having the ability to investigate your past. They can discover your skeletons via database searches on stories covered by the print and broadcast media. They also have access to information relative to inspections and fines levied against your company and your subcontractors by governmental agencies. Reporters take pride in uncovering this information and revealing it to the public at the most inopportune time. Well, that time has come, and you are in the hot seat.

Fingerpointing is not an option here because the findings for this particular accident will not be completed for some time. Therefore, it is important to stick to your key points like glue. However, when the findings are released, you need to be prepared to answer this question.

Possible Response: We will not know who or what caused the accident until the investigation has been completed. We cannot draw any conclusions until we fully understand every possible aspect of the situation.

Question: I understand that this project is behind schedule. Do you think this accident was caused by pushing your workers too hard to make up for lost time?

Discussion: Our hypothetical scenario mentioned that the project, located in Denver, was two weeks behind schedule due to heavy rains. The answer to the first part of the question must address the schedule issue head on. The second part of the question, however, addresses a rumor that the superintendent was pushing the workers to catch up. How did a reporter find this information? Rest assured that no matter how proactive you are, reporters seek interviews with anyone who has a heartbeat and the more emotional the person interviewed, the better. The more disgruntled the employee interviewed, the better. The more titillating the rumor, the better. Do not, repeat, *do not* participate in rumors or speculation. Be steadfast in your response.

Possible Response: Because of the heavy rainfall that Denver has experienced, we are approximately two weeks behind schedule. Let me be very clear in saying that XYZ would never compromise safety for schedule.

Question: We had a conversation with a lady who feels as though your workers have harassed her as she walks to work everyday. Could this accident have been caused by your workers paying more attention to the women walking by the site than to their jobs?

Discussion: Once again, this is a trap question that has the word *trouble* written all over it. Always bridge to your key points and repeat them ad nauseam.

Possible Response: We do not know the cause of the accident at this time. It would not be prudent to speculate about the cause until the authorities have had time to complete their investigation.

Question: Will you step up safety measures as a result of this accident?

Discussion: The implied assumption is that your safety program may have been lax prior to the incident. Make certain that you correct this assumption quickly and decisively. Also, you do not need to answer a question exactly as it was stated. Your audiences will pay more attention to your response than to the question.

Possible Response: XYZ has had, and will continue to have, a strong safety program that is enforced by a full-time safety director. There is no higher priority than the safety of our employees and those who work on the site.

Question: What are the estimated dollars in damage?

Discussion: This is a loaded question that is impossible to answer at this time.

Possible Response: That information will not be available until the extent of the damage has been determined. We will be happy to share that information with you as soon as it becomes available.

Question: Can we get in to take pictures?

Discussion: XYZ is responsible for ensuring safety on the site—no matter what. No one, especially nonpersonnel, is allowed on the site until it has been secured and cleared for entry by the authorities. Because our hypothetical crisis is less than 24 hours old, we can safely assume that nonpersonnel are not allowed on the site.

Once the site is secured and cleared for entry, however, XYZ may consider having its safety director or a member of the crisis management team provide a tour for one pool reporter and one pool cameraperson to show that you are willing to help with the story. (The word *pool* means that a single reporter and/or cameraperson goes on site to gather information related to a story and then share that information with other reporters.) Many of you are cringing right about now at the thought of letting two vultures onto the scene; however, reporters are going to get a story and pictures anyway, so why not consider being proactive and helping them do their jobs? You stand a much greater chance of controlling the content of the story if you provide the tour and dialogue. Prior to allowing the media on site, however, make certain they sign a hold harmless agreement and require them to don the appropriate safety gear. Safety must be enforced and consistently communicated in every way possible.

Possible Response: Once the site is secured, access will be determined by (name the authority). We will keep you advised of our progress.

John Smith should spend no more than five minutes answering questions because information will be scarce at this time. To terminate the interview, John should say, "I will take one more question, and then I need to return to the site." A reporter's closing question may be "Is there anything else I should know?" or "That is all of the questions I have. Is there anything else you would like to add?" This is your golden opportunity to reinforce your communication goals. Don't blow it—be prepared! John should use this opportunity to say, "There are no words to express our sadness for those involved in the accident. We don't know what happened, but we are going to find out and then do whatever it takes to make sure it never happens again." John should then announce the time of the next update and exit confidently.

Reporters are excellent ferrets for information and can gather it from a variety of sources:

- Employees who are leaving the job site.
- Employees who may have witnessed the accident.
- Background checks on XYZ and the mechanical subcontractor.
- Interviews with Mary Smith.
- Interviews with residents in the neighborhood.
- Interviews with XYZ employees at the corporate office or other job sites.

This is what reporters refer to as "finding balance" to a story, and you can do nothing to stop this activity. Now, let's talk about the *c*-word—control. You must understand the areas you can control and focus your energies on them, but you must also understand the actions and reactions that you cannot control. For instance, you can control your message(s) to the outside world, but if reporters are off your property, you cannot control who they talk to and the public information they can access. What you *can* do, however, is monitor the information that is being reported and prepare a response to that information during your next interview.

How can you provide the media with updates on your crisis?

As time progresses, additional information will become available for release. Be sure to release it before a reporter asks for it; otherwise, a certain reverse psychology will occur. If you present the media with information *before it is requested,* you take the bluster out of their questions because they did not have to pester you to obtain that information. Its level of importance may decrease in their eyes and so make them take a less adversarial approach toward your company. You can also build credibility with the media by submitting unsolicited information because doing so shows that you are proactive, are in control of the situation, and have nothing to hide.

The best vehicle for updating the media on your crisis is the news release, which is a one- to two-page document providing factual information on the crisis. The news release can be built upon the second foundation component discussed earlier in this chapter; this component organizes your information in the what, when, where, who, why, how format and includes a contact name and phone number for the reporter to call if he or she has any questions. Exhibit 5.1 shows an example of how a news release should be formatted.

Remember five points when developing a news release:

1. Keep it brief, but complete, by tightly organizing your information.
2. Deliver accurate information, such as correct spellings of names, full titles, time, date, location, and verified facts.

EXHIBIT 5.1 Sample News Release

XYZ Construction Company
123 Jasmine Street
Anywhere, USA

FOR IMMEDIATE RELEASE   **CONTACT: John Smith**
[Date]   [Phone]

WHAT	What happened?
WHEN	When did it happen [day/time]?
WHERE	Where did it happen?
WHO	To whom did it happen and who was involved?
WHY	Why did it occur?
HOW	What caused it to happen?
CLOSE	If human life was involved or the public was inconvenienced, express compassion and concern and state the current status of the crisis.

###

(*Source:* Janine Reid Group, Inc.)

3. Include the name, title, and phone number of a contact person.
4. Close your release with compassion or concern if human life was involved or the general public was inconvenienced. Also, make certain that you add the current status of the crisis and the steps the company is taking about the situation.
5. The three pound signs (###) at the end of the news release mark the end of the document.

A news release can be faxed or e-mailed to reporters as often as verifiable information is available and approved for release. Typically, the release invites phone calls from reporters, so the spokesperson must always be prepared to answer questions.

A statement release can also be used effectively during a crisis. For the sake of illustration, we can refer to the statement that John Smith delivered at 9:00 the morning after the accident. This statement can be converted into a written release and distributed to attending reporters as well as faxed or e-mailed to other interested parties. Exhibit 5.2 illustrates a sample statement release.

Furthermore, the company's employee base should receive a copy of each news/statement release that the company distributes. This notification keeps employees apprised of the progress of the crisis and the measures the company is taking to deal with it. Employee notifications also help to ensure that the same information is delivered to all of the company's audiences. The distribution of the release to the employees can be via e-mail or fax, or it can be tucked into the employee's payroll envelope, posted on the bulletin board, or communicated by supervisors. The goal is to communicate this information as quickly and consistently as possible.

EXHIBIT 5.2 Sample Statement Release

XYZ Construction Company
123 Jasmine Street
Anywhere, USA

FOR IMMEDIATE RELEASE   **CONTACT: John Smith**
[Date]   [Phone]

Miami, FL . . . XYZ Construction Company experienced an accident at 3:30 yesterday afternoon at the [name of the project]. A cooling tower was being lifted by a crane to the roof of the building when it broke loose at the eighth floor. We do not know the cause of the failure and are sorry to report that the accident resulted in three employees being injured. The status of the injured employees is as follows:

Juan Lopez is a laborer for ABC Mechanical Company and was released from the hospital last night. Juan has a broken leg and is expected to return to work within a few weeks.

Ron King is also a laborer for ABC Mechanical Company and is in fair condition at University Hospital. Ron has two broken ribs and a sprained ankle and is expected to make a full recovery.

Larry Lynn is a carpenter apprentice for XYZ Construction Company and remains in critical condition with a skull fracture. Larry is also at University Hospital.

We are staying in close contact with the workers, their doctors, and the families because their recovery is our primary concern. We will keep you updated about their progress.

Also, we are very concerned about determining the cause of this accident. We are working diligently with the authorities to determine the cause of the failure and will leave no stone unturned in reaching a resolution. We have called in an additional investigation team to make certain that all of the facts are known.

The site has been secured, debris has been removed, and we will reopen the project for work at 6:00 A.M. tomorrow. We will continue to update the media as verifiable information becomes available.

# # #

How do you handle media calls via the telephone?

John Smith has completed his interview and is returning to his office, where he will no doubt have a lot of phone messages waiting for him and more calls on hold screaming for his attention. Here are two suggestions that are extremely effective:

1. John should have all media calls screened by an assistant. This screening will buy him a few minutes to organize his thoughts and review his key points. Here is how the screening process works. John's assistant answers the call and says that he is on another line (or in a meeting) but will return the call as quickly as possible. The assistant then completes the media log sheet (Exhibit 5.3) and passes it on to John.

2. John then returns the call and records his statement on the media log sheet. A copy of the statement can be faxed to the reporter as a backup to the conversation. This backup helps eliminate any misinterpretation that may occur during the phone conversation.

EXHIBIT 5.3 Media Log Sheet (to be completed with each media call)

PUBLICATION/STATION _____

REPORTER _____

PHONE NUMBER _____ FAX _____

DATE/TIME OF CALL _____

DEADLINE _____

DATE/TIME CALL RETURNED _____

RESPONSE _____

(*Source:* Janine Reid Group, Inc.)

The media log sheet affords John an element of proof of what was communicated during the phone conversation. The media log sheet can also be used effectively as a follow-up for future interviews with the same reporter. For example, if John spoke with Sue Jones at Channel 4 yesterday at 10:30 A.M. and she is now calling for an update at 1:00 P.M., John might say, "Sue, when we spoke at 10:30 A.M. yesterday, the status was such and such. I am pleased to report that today . . . [blah, blah, blah]." Reporters will be taken aback by this technique because they are not accustomed to a spokesperson telling them what they were told in a previous interview. This technique can be very effective and can help minimize inaccurate reporting.

Is there a time when it is not in a company's best interest to talk to the media?

Absolutely. Let's say that your company is involved in a highly visible local project that is controversial on a number of levels and that has many parties involved. Let's also presume that the local media has become biased on the subject and is currently conducting a witch-hunt—with your company portrayed as the head witch. At this point, the crisis management team, along with upper management, must determine if anything can be gained by talking with reporters. If nothing can be gained and if the risks of damage to your company's public image are high in your participation with the media, then you should pass on any interview. In such a case, communicating directly with all of your audiences is the next step, which is discussed in Chapter 8.

What if you determine that you made a mistake after the interview is over?

Immediately call the reporter and say that you just realized that the information you just released was inaccurate and that you would like to submit an oral correction. Always follow up your correction in writing and submit it via fax or e-mail. A reporter wants to avoid inaccurate information as much as you do, so be honest and correct the situation as quickly as possible.

What are your communication obligations if you are a public company?

A crisis is a crisis—no matter what its shape or size; however, a layer of complexity is added if the crisis occurs to a publicly held company. Jay Kraker, managing director for Kraker & Company, which specializes in investor relations and marketing communications, has a great deal of expertise in this area. He is also the past director of corporate communications for Brown & Root, Inc. and the past managing editor for *ENR*. On this issue, Kraker offers the following comments:

> In times of crisis, fundamentally sound communications practices are identical regardless of a company's size, industry or location. But an extra set of rules applies to U.S. companies whose stock is traded publicly. These companies must comply with requirements of the Securities and Exchange Commission regarding disclosure of "material information." Although there are few hard and fast rules about exactly what is "material," any information that is likely to affect the price of a security should be considered material, as should information that would significantly alter the total mix of information currently available about a security.
>
> Public companies should consider disclosure requirements carefully when communicating about any crisis that has the potential to produce a significant impact on their financial performance. Failure to comply can lead to potentially significant financial liabilities if investors claim they did not receive access to the information in a timely fashion. Detailed guidance on disclosure issues and effective practices can be found in the *Standards of Practice for Investor Relations* published by the National Investor Relations Institute.
>
> Disclosure problems are most likely to occur in the rapid-fire decision-making that is inherent to crisis communication because such situations leave little time for discussion of disclosure issues. The cornerstone of compliance with federal disclosure rules is that companies must give all investors, large and small, equal access to material information. This requires that a public company disseminate material information widely by issuing a press release and distributing it broadly, using a recognized newswire service such as

PR Newswire or Business Wire. This must be done *before* the company provides this same information in news briefings, conference calls, or interviews where the audience is limited. In the era of instant, global communication via the Internet, it is important to note that a company does not satisfy disclosure obligations purely by issuing a press release on its home page on the World Wide Web.

As always, the first few hours of a crisis are the most critical. Managers far from headquarters who temporarily assume communication responsibilities during this time are rarely as versed in these requirements as the director of investor relations or corporate communications. Similarly, employees of a privately-held firm may not even be aware that their customer must comply with disclosure rules. These represent just a few of the possibilities that underscore the importance of another bedrock principle of effective crisis communication: having a single person speak for the company. (Personal communication, June 28, 1999)

What are some guidelines for the spokesperson to follow?

- Establish your company as *the* source for information with the news media at the outset of a crisis. If you fail to do so, reporters will find other sources who will be more than willing to talk to them. Always advise reporters on the time and location of the next update.

- Develop *your* agenda prior to any interview. In other words, determine what your communication goal is, have a firm grounding in your facts, and focus on what you can say—not on what you cannot say. To control your side of the story, you must be prepared; otherwise, you will be perceived as unorganized and confused and so lose credibility.

- Be accessible to reporters—even if there is no new information to report. If you do not have an answer to a reporter's question, say so and then explain why. Perhaps it is information that you can gather and report to him or her shortly; maybe it is information that you will not know until the investigation is complete. Whatever the case, do not be afraid to say, "I don't know," but then follow that statement up with "but I will find out and get back to you by (time)."

- Reporters may use a technique called *rapid-fire questioning,* where they throw many questions at you without pause. This technique can be unnerving, even for veteran spokespeople. Should rapid fire occur, simply say, "I will be happy to answer all of your questions. Let me start with. . . ." Then pick the question you wish to answer and go from there. As discussed earlier in this chapter, your answer will be remembered by the reporters' readers, viewers, and/or listeners more than the question itself.

- Use examples, anecdotes, or visuals whenever possible. Both newspapers and television are highly visual media, and reporters will most likely use any examples that you can provide. Thus, give them photographs, drawings, charts, or graphs. Also, use comparisons with familiar things. For example, if you experienced a spill that covered 40,000 square feet, you might say, "The affected area is 40,000 square feet, which is the approximate size of a football field." Such a comparison allows readers/viewers/listeners to grasp the size of an area quickly. Once again, the focus is on communicating your message rapidly and succinctly.

- Be aware of a reporter giving you the silent treatment. Here is an example: John Smith has just finished delivering an answer to a reporter's question. The reporter looks at John, keeps the microphone at John's mouth, and says absolutely nothing. What will John do? I'll tell you what he will do. He will do what any of us would do—fall into the abyss. That microphone will become a verbal vacuum cleaner and suck every wrong word out of John's body. Reporters have been using this technique for decades. It works! It works with live interviews and telephone interviews. Here is a tip that will help you through this uncomfortable time. When you are finished with your answer, stop talking—no matter what. If the reporter starts whistling the Jeopardy tune, simply say, "What is your next question?"

- Develop a fact sheet that includes pertinent information relative to your company and the crisis at hand. The fact sheet can serve as a support for your spokesperson as well as a handout for reporters.

- A reporter has control over the editorial content of your story—you do not. A reporter's job is to ask questions and develop a story. Your job is to anticipate those questions and respond in the most honest and favorable way possible.

- Never let your guard down and always assume that the microphone is hot and that the camera is rolling. You can relax only when you see the skid marks of the reporter's car leaving your facility—but not until then!

- Tape recording your interview is perfectly acceptable, as long as you notify the reporter that you are doing so. Recording your interview serves two purposes: (1) it puts the reporter on notice that you have a record of what is being said, a step that may increase the likelihood of your statements being reported accurately, and (2) you can use the recording as a learning tool to help you with future interviews.

- Seek media training for your spokesperson. Working with the news media during a crisis is showtime—not on-the-job training. If a spokesperson is not prepared for the pressure, a slip-up may occur that causes the company's reputation irreparable damage.

- Consider the old warning "Never argue with anyone who buys ink by the barrel." In other words, never argue with a reporter, because you will never win. A reporter's provoking questions probably will not be reported, but your angry reply will. Therefore, stick to your key points and mantra.

- Never repeat a negative question. For example, if you have an accident that resulted in a fatality, a reporter might try to trick you with the question, "Do you always kill people on your projects?" Do not repeat the negative and reply with "No, we do not always kill people on our projects." The implication is that you kill people on only a few of your jobs.

- If a reporter's question contains incorrect information or facts, do not be afraid to challenge the reporter and correct the information immediately. If you do not take care of misinformation, it will haunt you throughout the crisis and beyond.

- Never, ever, ever, say, "No comment." Period. This remark is analogous to saying, "I am so guilty that I can't even stand myself!" Sometimes, however, when information is sensitive or the situation is in litigation, it is inappropriate to say anything. Please, take the sting out of those two words and explain why you cannot comment. An example might be "The situation is in litigation and it would not be appropriate for me to say anything at this time."
- Avoid the use of industry jargon and speak in common, everyday English that John Q. Public can understand.
- Return reporters' phone calls quickly and pay attention to their deadlines. If you do not contact them prior to their deadlines, you might read or hear "The company had no comment" or "The company could not be reached for comment." Call them back even if you do not have any new information.
- Always be willing to learn from other spokespersons. When a crisis breaks in your local area, follow it through to its conclusion and study the techniques used by the spokesperson. You may pick up some great tips as well as learn areas to avoid.

Never doubt the power of the media to manipulate public opinion. A mobilized public can, and has, shut down many companies that did not respect this power. Just watch or read the news on any given day for a confirmation of this fact.

Chapter Summary

- Many construction incidents have resulted in major media events because of slow or nonexistent responses from the companies under siege. Such responses can cause irreparable damage to a company's reputation. Bad news does not improve with age, so a company's fast response to media interest is vital.
- A buy-time statement should be delivered by the temporary spokesperson at the outset of a crisis. This statement establishes the company as proactive with the media, yet it allows the company time to gather information and get the corporate spokesperson to the site. Avoid answering questions after the buy-time statement because little verifiable information will be available.
- Your statement(s) should be built upon a solid structure of a beginning, key points, and an end. One of your key points must communicate concern and compassion for anyone who has been hurt, killed, or inconvenienced as a result of the incident.
- A successful outcome to a question and answer session with a reporter depends on anticipating questions that may be asked, by developing responses for those questions, and by rehearsing the competent and confident delivery of those responses.
- Never send a spokesperson into a question and answer session with a reporter without a mantra—that is, a rehearsed answer to apply to difficult questions.
- The bridging technique enables you to transition from a difficult question to one of your key points. In a question and answer session, this technique is invaluable

because it takes you from where you are to where you want to be. Always look for opportunities to bridge to your key points to ensure that your message or messages are communicated loud and clear.

• You are not obligated to answer a question *exactly* the way it is presented to you. Just listen to any politician if you doubt this advice. Do not hesitate to reshape a question and bridge your response to one of your key points.

• Always, always, always express concern and compassion for anyone hurt or inconvenienced as a result of the crisis.

• A company can, if it wishes, refer all media calls within the first few hours of a crisis to the authorities on site. This referral is appropriate only for certain types of crises, such as natural disasters, workplace violence, sabotage, and so forth. Such a referral, however, does not protect a company from the media for the duration of the crisis, so preparation and organization is still required.

• Saying the words "I don't know" does not imply that you are incompetent as long as you follow them up with "but I will find out and get back to you by (time)." The same technique applies if a reporter blindsides you with, say, a rumor. Simply say, "I am unaware of that, but I will investigate the situation and get back to you by (time)."

• Anticipate as many questions as possible before an interview with a reporter. This preparation gives you some time to rehearse your answers and determine how to bridge to your key points and mantra.

• Stick to your communication goal and repeat your key points at every opportunity. Do not forget, to be brief in your answers.

• All media telephone calls should be screened by the spokesperson's assistant. The spokesperson must have a few minutes to organize his or her thoughts and key points.

• Do not ask reporters to see their stories before they are printed or aired. Once the reporters stop laughing, they will politely tell you that it is not possible, so save yourself the embarrassment.

• Monitor and record news coverage that involves your company. Do not be blindsided by problematic coverage that you could have spotted and prepared responses for.

6

MORE ON THE GOOD, THE BAD, AND THE TRULY UGLY

What types of media will cover your crisis?

In previous chapters, the news media has been referred to as a single entity; however, it encompasses five areas:

1. Television
2. Radio
3. Newspapers/Magazines
4. Trade publications
5. Internet

The needs of each of these media outlets can, and do, vary; therefore, a general understanding of their individual needs will help you communicate your messages more effectively during a crisis. Here is a review of the five types of media and tips on how to work with each effectively.

Television

The words *television* and *immediacy* go hand in hand. Television is driven by the constant pressure of deadlines coupled with the insatiable need for fast information and visual stimulation. The stories reported on television are typically short and rely heavily on visuals. Remember the old saying "A picture is worth a thousand words?" Television journalists live by it; they recognize that a visual image depicting mayhem and destruction can tell a story in mere seconds. More importantly, from a construction company's point of view, those images can negate or dilute the words of its spokesperson in a crisis.

The immediacy factor that drives television reporting can create another problem for a company in crisis. Much like oil and water, speed and accuracy do not mix. As discussed in Chapter 3, you must consistently monitor and videotape all coverage of your crisis. If a reporter did not get the facts correct or misinterpreted any of the information communicated by your spokesperson, an immediate correction should be issued or communicated in your next statement. This countermeasure for misinformation is discussed later in this chapter.

Videotaping your spokesperson also gives you the opportunity to critique his or her ability to communicate the company's messages effectively and to answer questions succinctly. These tapes can also serve as excellent training tools during your post crisis evaluation process, which is discussed in Chapter 11.

Monitoring services are available to assist you in the process of tracking media coverage. Consult the yellow pages of your telephone book or locate a service via your public relations counsel.

As we discussed in Chapter 5, there is no substitute for preparation. Typically, you are on a short fuse to prepare for media interviews during a crisis; however, take time to collect your thoughts, organize your key points, and anticipate questions. Chapter 5 reviews the basics for getting ready for an interview with a reporter, but here are a few additional tips for a face-to-face television interview.

1. When television reporters arrive at your site to cover your crisis, they are each accompanied by a camera person. Attached to the camera is a light strong enough to illuminate an airport at midnight. This glare is an unsettling experience, even for the pros. The best way to prepare for it is to conduct a mock interview with your spokesperson using similar equipment. Rehearsal reduces the shock of the bright light and lessens the anxiety it can cause. Clearly, this type of training is most useful prior to the outset of a crisis.

2. The company's spokesperson should always focus on eye contact with the reporter—not the camera or any activity in the area.

3. Dress conservatively. Men should avoid wild patterns on shirts and ties; women, on blouses and dresses. Also, women should avoid distracting jewelry, such as dangling earrings and very large necklaces.

4. To avoid looking like a wooden figure, use gestures to illustrate points and nod your head to acknowledge the reporter's comments; however, avoid touching your face, playing with your hair, or constantly adjusting your glasses. These types of gestures communicate nervous energy and dilute your persuasion and credibility.

As discussed in Chapter 4, watch spokespersons from other companies/organizations who are being interviewed on local/national news or news magazine shows. This study helps you learn how to handle televised interviews.

Television interviews are either live or taped. For the prepared spokesperson, a live interview is more effective because the story is running live and cannot be edited. In contrast, a taped interview can be edited, and much of what the spokesperson said can be deleted or edited out of sequence—yet another reason to rehearse and to know your key points inside and out!

Radio

Similar to television, radio is driven by deadlines and the need for fast information. Radio newscasts are aired more frequently than television newscasts. Radio interviews are conducted in both live and taped formats. Many of the taped interviews occur via telephone and require the same amount of preparation as face-to-face interviews. Do not forget to practice the art of brevity, because radio news stories are typically very short. The exception occurs when your crisis becomes the focus of a talk radio show.

When a crisis strikes, a radio station will most likely discover your crisis through television or newspapers. This coverage is yet one more reason why it is imperative to monitor those media during a crisis.

Newspapers/magazines

The print side of the media equation is not subjected to the live reporting intensities that their broadcast relatives experience. On balance, a print reporter may have a bit more time to develop a story and, therefore, contribute more substance to that story. Frequently, you can hear a broadcast reporter/anchor parroting a story covered in the morning or evening paper. The media feed off one another's material.

Trade publications

Of all of the media described above, the trade press delivers the most balanced coverage of crises, for two reasons:

1. Your trade press understands your industry and your industry sustains your trade press. Sensationalizing your incident and pronouncing you guilty until proven innocent is not in the best interest of your trade press. Its job, however, is to report the crisis and share verifiable information with its readers in a balanced fashion.
2. The large majority of the industry trade press is published on a weekly, bimonthly, or monthly basis, so additional time can be allocated to cover the story in depth.

Even though the trade press has a better understanding of the construction industry than the general press does, you must still be diligent in your preparation and response to its requests for information.

Internet

Writing about the Internet as a news medium is difficult because it relies on an ever-changing technology and because the time may still too early to understand the full implications of the Internet as a news source. Nevertheless, it is a medium that should not be ignored. Currently, the Internet press consists of trade-focused Web sites as well as major publications and networks that feature breaking news Web sites.

As you can see, the five media types described above have a few differences and many similarities in the ways they gather and report news. Keep in mind that regardless of the medium—electronic or print—the reporters' goal is to tell a story, so their questions will be the same. Your goal is to remain consistent with each statement, no matter what media type you are addressing. To maintain this consistency, you must be prepared. No magic pill can be taken to enhance the outcome of an interview. The key is preparation.

What if the media tries to get information from your employees?

Let's return to the XYZ Construction Company crisis. It is 4:30 p.m. on the day of the accident, and John Smith has just delivered his statement to the media. The decision has been made to suspend work until further notice. The witnesses are still in the job-site trailer being debriefed, and the rest of the workers have been released to go home. Well, guess who is waiting in the parking lot (or the local taverns) to talk to the workers? And guess who is waiting in the parking lot at the corporate office of XYZ Construction Company? Good guess—reporters!

Reporters gather information from any source they can find, and employees can provide excellent fodder for their stories. Reporters like to focus on people in a highly charged emotional state or, even better, a disgruntled employee. Reporters' jobs are made easy if they can catch employees walking to their vehicles from the crisis scene and begin the incessant questioning process. If reporters cannot catch a "victim" this way, they may follow the employee home and try to initiate an interview there.

So how do you counsel your employees to prepare for such a situation? First, all employees should be warned that the media may be waiting for them as they exit the site/building. Second, instructions should be given on what employees should say if a media encounter takes place. One suggestion is, "I'm sorry, I don't have the information you need, but John Smith can answer your questions, and you can reach him at (phone number)." Finally, employees should be instructed to immediately walk away after delivering that comment and not entertain any questions.

This technique works in many situations; however, reality rears its ugly head when employees elect to use this opportunity as a soapbox for their gripes. Employees may also elect to call their local radio or television station and leak information on the company's crisis in order to win a "Hot Tip" money award. Cyberspace is another venue that employees may use to air their grievances.

At this point, you must recognize what you can and cannot control. Once employees are off your property, you have no control over what they say or do. However, you can monitor what is being said and prepare a response to defuse the situation. You can also ensure that the flow of information from management to employees is consistent from the outset of the crisis to its conclusion. A good way to ensure consistent communication is to distribute a copy of all news and statement releases to employees so that everyone stays current on the situation. Keep in mind that no matter what herculean efforts you make, the rumor mill will still be in full swing and leaks will occur. However, if you stay consistent with employee communications, you will generate a great deal of loyalty among the largest percentage of your employee base.

How do you handle a hostile or pushy reporter?

Most reporters conduct themselves professionally to get the best possible story; however, a feeding frenzy can occur when a crisis strikes. This reaction is typically due to the highly competitive nature of the media industry as well as to the lack of available information during the first few hours of a crisis. Some reporters become hostile when they believe that they are being stonewalled by the company spokesperson. To prevent this situation, here are pointers to help your company's spokesperson display grace under pressure.

1. Be prepared by knowing all of the important facts and details that have been verified and approved for release.
2. Focus on what you can say versus what you cannot say.
3. When the reporter asks you a question, give yourself time to think by pausing for one or two seconds before responding—or restate the question in your own words.
4. If a reporter delivers a series of rapid-fire questions, pick the one that best relates to the point you wish to make.
5. Maintain your position of authority by making eye contact with the reporter and communicating confidence in your statements.
6. If a reporter interrupts you while you are making a statement, wait until the reporter is silent before completing your statement and continuing with any other points you want to cover.
7. Here is the toughest one. Keep your cool—no matter what!

As I noted in Chapter 2, reporters have been known to wander into off-limits areas during post accident tours. Worse yet, reporters can engage in deception to get their stories. Over the years, I have witnessed some shameless reporting techniques. For example, one reporter obtained the overalls of a firefighter, along with a hat and badge, to gain access to a site so he could interview the witnesses. Once discovered, he became indignant and refused to leave the site. Another reporter entered a site by hanging from the undercarriage of a fire truck!

If reporters gain unauthorized access to your site, you not only have a safety issue but a trespassing issue as well, and their removal should be immediate. The authorities (fire/police) should be used to accomplish this task because their greater power helps you to avoid a potential conflict.

The creative efforts exhibited by reporters do not stop at the many disguises they take on. For instance, the driver of a ready-mix truck was proceeding through an intersection when four intoxicated teenagers ran a red light and collided with him. All of the teenagers were killed. The ready-mix truck driver had minor physical injuries but major psychological trauma, which kept him in counseling for six months. A local television station's investigative reporter, who covered the initial accident, decided to breathe life back into the story by interviewing the driver on a monthly basis. She called his home and begged the family to put him on the phone for an interview. When that did not work, she bought presents for his three children, thinking that would soften him up. This is blatant harassment. In such a case, if the reporter calls again, a family member should state that the driver is having a difficult time dealing with the tragedy of the accident and that he would appreciate not being called again. If this effort fails, the family should contact the station's management and file a complaint, in both oral and written forms.

Many companies in the construction industry maintain solid media relations programs. They mistakenly believe such efforts will insulate them from hostile reporters. However, while a proactive media relations program is important (see Chapter 9), quite possibly the reporter you have a relationship with will not be the one to cover your crisis. Also, if you are the culprit in a crisis, no one is going to cut you any slack—no matter what relationships you have developed over the years.

Before we close this section, focus on tip number seven, which says, "Keep your cool—no matter what!" This advice is a challenge because, generally, the media employs a double standard. On one hand, reporters can ask provoking questions to get particular responses from company spokespeople. On the other hand, if the spokespeople become defensive as a result of the questioning, they have their heads handed to them in broadcasts and print. Thus, you must rehearse keeping cool. Have colleagues pepper you with hostile and provoking questions, and you practice keeping cool.

Also, if reporters misquote the spokesperson, take comments out of context, or cite rumors or speculation, their stories can create a tremendous amount of damage that a company will have a difficult time correcting. Have some irresponsible reporters spoiled the bunch? In a word, yes. But we have no choice but to learn how to work with them and defend ourselves through preparation and effective delivery of verifiable information. The next section, as well as Chapter 8, pursues this topic in more depth.

What if the media misrepresents the facts, takes your statements out of context, or misquotes you?

The time may come when you will scream, "I will never speak to another reporter for the rest of my life. No matter what I say, they screw it up!" This has become the theme song for many a company spokesperson, and it will remain on the Top Ten list for a long time to come.

Two possible reasons explain misrepresented facts or your company spokesperson being misquoted or taken out of context. One reason focuses on the media's side; the other, on the company spokesperson's responsibilities. Let's take a look at the media's side first, as they are easier to pick on.

Patricia Raybon, an associate professor at the University of Colorado at Boulder and a former Denver newspaper journalist, lists the following as reasons for journalists sometimes getting their facts wrong:

We're in a hurry. Deadlines in daily journalism prompt reporters to take shortcuts. Eyes on the clock, we cut corners, often taking cursory looks at otherwise critical information. So often we get our facts wrong or incomplete.

We repeat each other's errors. Reporters often cite information from published stories stored in news data bases. This saves time, but reporters can't be sure that the original stories were accurate as printed. Thus, errors get recycled.

We wear blinders. Looking with fresh eyes at an issue takes work. So often we are less likely, as journalists, to challenge our own longstanding beliefs. What results is reporting by cliché. Often, it's far off the mark.

We rely on the same sources, again and again. It's a recycling problem that narrows the number of participants in important public debates. Fresh voices add fresh air. Most newsrooms need more.

Reporters use each other's friends as sources. I can't count the times I've heard a reporter turn to nearby desks and ask colleagues: "Anybody know anybody who can give me quotes on such and such?" I've done it myself. This insular arrangement excludes new, potentially insightful people from news dialogue.

We're lazy, and the clock is always ticking. Good journalism takes energy, hard work, enthusiasm, and courage, not to mention skill. While legions of journalists work hard everyday, often with low pay, too many of us are no longer motivated to give our readers our best. So good stories get overlooked, or underreported.

We're biased. Working off a hunch or, indeed, off a bias isn't inherently bad. But at the end of the day, both sides of an issue are supposed to be reported, not just the side the reporter favors. In public-affairs journalism especially, a reporter's bias is a place to start an investigation—not the place to finish it.

It's too easy to find a legitimate source who will simply repeat a reporter's personal views. That then becomes the news. That is a disturbing practice in American newsrooms, one that abuses [a reporter's] privilege. (Personal communication, July, 1999)

Ms. Raybon's candid analysis of the news media is refreshingly on target. However, it does not give a company permission to lay blame on others and justify its actions. Misrepresented facts, out-of-context quotes, or misquotes can also be the fault of the company spokesperson. Many times these errors occur because the company spokesperson was unprepared and delivered an unorganized and disjointed response to a reporter's questions.

So the question arises, "Are both parties equally guilty?" Probably, but the media holds the trump card because they have the power to capitalize on your misfortune.

Having worked with the news media on scores of crises, I can honestly say that there are opportunities to turn a negative situation around; however, if you feel that your efforts have created only break-even coverage but your reputation and credibility are intact, consider it a victory. It could be much worse. See Chapters 11 and 12 for ideas relative to crisis recovery.

Be realistic about the pursuit of "exact" reporting and understand just how little control we have over what the media hears and therefore writes or says. In *The Pursuit of WOW,* Tom Peters (1994) offers interesting advice: "Don't get worked up about out-of-context quotes. All of life is 'out-of-context.' Face it, the only story that will make you happy begins, 'The brilliant Martin Mainman gave me an hour of his precious time last Thursday, and here's what he said . . . (a full transcript follows).' Guess what? It ain't gonna happen" (p. 285).

Probably a time will come when you feel that the media has misrepresented your situation or that your comments have been misquoted or taken out of context. Steve Gray, a veteran TV reporter and president of BVP Media, Inc., in Denver, Colorado, offers these words of advice from a broadcast perspective:

> There is no bigger surprise than hearing your words in a broadcast news story and those words have absolutely no resemblance to what you recall saying. This is a frustrating experience, but here are some things to consider should this happen to you.
>
> For the sake of illustration, let's assume that you have done all of the right things in preparing for your interview with a reporter and let's assume that you have said all of the right things. Let's go one step further and assume that the reporter has it all wrong. When the story airs, you are obviously very surprised. What can you do?
>
> First, use common sense. If the misstatements of fact are not central to the story, it may be best to let sleeping dogs lie. Remember, to correct an error you must repeat the error back to the audience in order to state the facts correctly. It may be more damaging to restate the facts than to have them corrected on the air. However, if the facts are central to the story and your company's position on that story, it is essential to correct the misinformation.
>
> There are two things that you need to keep in mind when you attempt to correct misinformation:
>
> 1. Always start with the reporter. There will be a temptation to go over the reporter's head to his or her editor, producer, or news director but this is not a productive move. It violates basic newsroom protocol, and more importantly, it is likely to agitate the reporter. The exception to this rule is if the news anchor misquotes or misrepresents your position. In this case, it may be appropriate to address your concerns to the show producer or writer.
> 2. Keep emotions out of your conversation with the reporter. This will be difficult, especially if you feel as though you have been mistreated, but it is vital to successful media relations. When you initiate your call to the reporter you can simply say, "During the course of our interview, I said X and you reported Y. Would you please explain how that could have happened?" This will give the reporter a chance to tell their side of the story. If you do not receive satisfaction from the reporter, it is acceptable to move up the chain of command. Always

remember to be swift and firm in your response and always stick to the facts. (Personal communication, July 1999)

Over the years, I have seen too many companies fail to stand up for themselves when they have been misquoted or taken out of context. Whether it stems from fear of reprisal or from wanting to put the matter to rest, such reticence is a mistake for two reasons:

1. The misinformation will reside in a publication's/station's database (as well as other databases) for a very long time and will resurface the next time your company enters the media's spotlight.
2. The reporters may continue to be lax in their responsibilities and print/broadcast additional inaccuracies.

Reporters are not perfect, and companies must stand up for themselves with the media if they feel they have been unjustly harmed. A word of caution applies here, however. You must be absolutely positive about what transpired in the interview before you make your challenge. Once you lodge your complaint, you may receive a correction or you may not, but it is certainly worth your time and effort to stand up for yourself.

Now here is a concept that will probably get a cool reception, but it is worth discussion. If you believe that a reporter did a good job of reporting your story, consider calling the reporter to thank him or her for spending the time to acquire and communicate the known facts in a fair and balanced fashion. This thanks may be the last thing you would think of offering in the heat of a crisis, but it can—and does—go a long way toward promoting positive media relations. If you cannot do so during the crisis, wait until the situation is resolved to thank the reporter.

What if you are guilty?

As we discussed earlier, a company in crisis is judged more on its response to the crisis than the actual crisis event. This tendency supports the main thesis of this book, which is that you must be prepared for any type of situation, whether you are the victim or the culprit.

If your company was at fault in a crisis, you must step up to the microphone, admit the problem, and communicate what you are doing about it. In the vast majority of crises where a company/person is found guilty, contrition has proved the most effective way to put the negative situation to rest and move forward. By exposing the negative facts as quickly as possible, you avoid letting the news media uncover them and dribble them out—day after endless day.

The general public is conditioned to expect defensiveness and even cover-ups from companies in crisis. But what they truly want, and crave hearing, is that people are doing something about the situation and will take their lumps if need be. Your audiences and the general public can forgive an admitted mistake, but they will not forgive a cover-up or blatant lying. A good rule to follow is "If the truth hurts, tell it all and tell it fast," because your credibility is at stake.

In "Regaining Credibility," James Lukaszewski (1999), chairman of The Lukaszewski Group, Inc., offers powerful insight on the topic of credibility in adversity:

> The toughest part about credibility is recognizing that you don't have it and that you can't declare it. Credibility is conferred on you, your company, your products, and your issues by those directly affected. Cleverly worded brochures, news releases, and speeches are not the foundation on which to build credibility.
>
> Behavior, based on fundamentally sound decision making and positive actions from the public perspective, is what leads to credibility. Even when things go wrong, there is a seven-step pattern of behaviors which can help obtain public forgiveness and rebuild credibility. Those steps are:
>
> 1. Candor. Outward recognition, through promptly verbalized public acknowledgment (or outright apology), that a problem exists; that people or groups of people, the environment, or the public trust is affected; and that something will be done to remediate the situation.
>
> 2. Explanation (no matter how silly, stupid, or embarrassing the problem-causing error was). Promptly and briefly explain why the problem occurred and the known underlying reasons or behavior which led to the situation (even if you have only partial early information). Also talk about what you learned from the situation and how it will influence your future behavior. Unconditionally commit to regularly report additional information until it is all out, or until no public interest remains.
>
> 3. Declaration. A public commitment and discussion of specific, positive steps to be taken to *conclusively* address the issues and resolve the situation.
>
> 4. Contrition. The continuing verbalization of regret, empathy, sympathy, even embarrassment. Take appropriate responsibility for having allowed the situation to occur in the first place, whether by omission, commission, accident, or negligence.
>
> 5. Consultation. Promptly ask for help and counsel from "victims," government, and from the community of origin—even from your opponents. Directly involve and request the participation of those most directly affected to help develop more permanent solutions, more acceptable behaviors, and to design principles and approaches which will preclude similar problems from re-occurring.
>
> 6. Commitment. Publicly set your goals at zero. Zero errors, zero defects, zero dumb decisions, zero problems. Publicly promise that to the best of your ability situations like this will never occur again.
>
> 7. Restitution. Find a way to quickly pay the price. Make or require restitution. Go beyond community and victim expectations, and what would be required under normal circumstances to remediate the problem. Adverse situations remediated quickly cost a lot less and are controversial for much shorter periods of time.

Numbers one and six are generally the toughest to accept. But, which would you rather believe and trust, the organization that says, "There's a little risk with everything . . . ," or the one that says, "We're committed to achieving zero accidents?" The biggest benefit of following this model is that it controls and sometimes dramatically reduces the tough follow-up questions that the media, along with your audiences, will ask of you.

When something bad occurs, organizations lose their credibility because they do the exact opposite of these seven steps. They deny. They obfuscate. They are arrogant. They stall. They delay. They avoid responsibility. They threaten to sue. The reality is that organizations needing to retain their credibility will carry out each of the seven steps in some order (some times with great pain) as part of their reputation recovery process.

One last word: A company clearly does not need to admit guilt if there is none to admit, but it does need to show that it cares and offers to do whatever it takes to restore confidence and credibility in the company.

How do reporters perceive themselves?

Over the last few chapters, we talked about a variety of issues related to the news media. We reviewed how reporters do their jobs in preparing a story and how a spokesperson can assist in that process while still watching out for the company's best interests. We also talked about the various techniques reporters use to get information. The best way to close this chapter is by sharing research from the Pew Research Center on how the news media view themselves and the job they are doing.

The Pew Research Center for the People and the Press, located in Washington, D.C., is an independent opinion research group that studies attitudes toward the press, politics, and public policy issues. Pew (1999), in association with the Committee of Concerned Journalists, conducted a survey of 552 top executives, mid level editors and producers, and working reporters and editors from both the national and local news media to determine the top issues facing the news media. The survey results were finalized in February 1999, and here are the highlights:

- Journalists increasingly agree with public criticism of their profession and quality of their work. Overwhelmingly, news media professionals say the lines have blurred between commentary and reporting and between entertainment and news. A growing number of reporters, editors and news executives also say that news reports are full of factual errors and sloppy reporting.
- Across all media—print, television, radio, and the Internet—the news media share public misgivings about their watchdog role. More of the news media today, than just a few years ago, say the press drives controversies rather than just reports the news.
- Lack of credibility is the single issue most often cited by the news media as the most important problem facing journalism today.
- To reporters and editors, the reasons for journalism's problems are clear—growing financial and business pressures. At both the local and national levels,

majorities of working journalists say that increased bottom-line pressure is hurting the quality of coverage.

- Most journalists and news executives agree that they are overly focused on internal dynamics, too often competing with one another and writing more for colleagues than consumers.

- The press gives itself only lukewarm grades for striking the right balance between what audiences want to know and what is important for them to know. This criticism is especially strong in the national television news; only 38% of those working in this arena say an appropriate balance is struck today, compared to 60% of the national print press and more than half of the local news media in both mediums.

- Two thirds of those in national and local news say that news organizations' attempts to attract readers or viewers has pushed them toward infotainment instead of news.

Can 552 representatives of the news media communicate a consensus that is felt by all of their colleagues? It is my opinion that most of the general population would concur with the Center's findings. This research demonstrates that we need to be vigilant in our quest to learn how to work with the news media effectively and defend our position if we feel we have been treated unfairly. In any event, the media's self-image should have absolutely no bearing on a company's need to be prepared for the inevitable interview.

Chapter Summary

- Technology is developing at warp speed; therefore, the news media can communicate your bad news instantaneously. This capability places a demand on companies to respond as quickly.

- Always monitor and record the broadcast and print media attention your company receives during a crisis. A review of all interviews should be conducted as soon as possible so positioning strategies can be determined for the next set of interviews.

- Consistent communication with a company's employees during a crisis is paramount because they are a company's voice to its external audiences. Employees are one of your most important audiences when a crisis strikes, so consistent communication with them during a crisis should be a priority.

- When a crisis occurs, many situations develop that you cannot control, such as conversations that your employees have with reporters. Do not spend a lot of time worrying about things you cannot control. Instead, focus on what you *can* control and monitor the rest so you can develop a position if you or your company needs to respond to negative comments from an employee.

- If a reporter violates your request to remain in a safety area, ask the on-site authorities to handle the situation. An authority figure is more effective, and you need not be involved in a potentially uncomfortable situation.

- Reporters have complete control over the content of their stories. The only control you have concerns the words coming out of your mouth. However, remember that reporters are human and that they make mistakes. If you feel you have been mistreated or misquoted by a member of the news media, make certain that you address the issue with the reporter. If you are not satisfied with your conversation, contact the reporter's superiors.

- If you or your company is at fault in a crisis, do not run and hide, nor should you lay blame on others and justify yourself. You will garner far more respect and credibility if you admit to the situation and communicate what your company is doing to make certain that it never happens again.

- Look for opportunities to learn from other spokespersons and companies in crises. Experience, especially someone else's, is a fabulous teacher. Also, learn from each interview that your spokesperson delivers and apply this knowledge to subsequent interviews.

- If you feel that a reporter did a good job in covering the story, try to find the time to thank him or her for taking the time to do a fair and balanced job. This effort will go a long way toward establishing strong media relations.

7

NEWS CONFERENCES

When is a news conference advantageous?

In general, you should call a news conference only when your organization is releasing a statement, responding to breaking news, or when an important announcement needs to be disseminated immediately to a wide range of media. Otherwise, there is no advantage in holding a news conference. They can be both costly and embarrassing because reporters hate attending news conferences that they feel are unnecessary. Conversely, not holding a news conference when it is clearly needed also sends the wrong signal. Therefore, it is in your best interest to know when a news conference should be held.

Unfortunately for some readers of this book, news conferences will be necessary simply because their crises have all the elements that the media seems to crave: conflict, drama, and shock value, among others. Thus, when a situation contains these elements and has an impact on the local community or poses a threat to that community, it will be considered a newsworthy story.

Accidents that result in injuries or fatalities generally put the media at your doorstep. Moreover, if the crisis occurs on a highly visible project, such as the construction of a new football stadium or at a site in or near a major metropolitan area, the media is interested in the story. In these situations, the media demands immediate information from your company, so it is clearly advantageous to hold a news conference.

For illustration purposes, let's return to our hypothetical situation. XYZ's corporate spokesperson, John Smith, has just arrived at the scene of the accident. A buy-time statement has been issued and an update has been provided. If Smith were faced with a swarm of reporters demanding additional information, coupled with a log of phone calls from the press, it would clearly be to his advantage to schedule a news conference. This would allow him the opportunity to disseminate information to a group of reporters at the same time rather than conduct a multitude of one-on-one interviews. If, on the other hand, the media has shown only moderate interest in the situation, no news conference is needed.

Whether or not your story attracts media coverage often depends on the other stories that are hitting at the same time. If it is a slow news day—that is, there are no major stories hitting—your situation may attract significant media attention, requiring a company response. However, if the media does not show interest in your situation, do not change that good fortune by holding a news conference. In summary, hold a news conference if:

- The crisis is likely to attract significant media coverage. As mentioned in Chapter 4, if you experience a situation that is leaking, flaming, exploding, hurting, dying, obstructing, delaying, decaying, polluting, or falling, it will most likely draw significant media coverage.
- The crisis poses a threat to the surrounding community.

A news conference is unnecessary if:

- The crisis has not attracted a great deal of media attention and the spokesperson can handle the requests for information via telephone or news release.

Where should you hold a news conference?

As I emphasize throughout this book, preparation is vital. Do not compound a crisis by holding a poorly planned event because it will not reflect well on you or your organization. In short, this is no time to wing it. Too often, companies unprepared for the rigors of news conferences are accused of sending mixed signals or conflicting messages.

Unfortunately, if you are holding a news conference, it probably means that you are responding to breaking news, so you will not have a great deal of preparation time. Your first step is to select a location. Choose a neutral site, such as a hotel conference room, rather than the scene of a crisis or your corporate office. Too many distractions inhabit the scene of the crisis, and your corporate office typically provides reporters with an opportunity to wander around and talk to employees.

When selecting a facility, ask yourself the following questions before settling on the location. If you cannot answer *yes* to the these questions, the needs of the media will not be met.

- Does the facility have adequate parking? Reporters and camera people can get cranky if they have to lug heavy equipment for blocks to get to a news conference. The last thing you want is a cranky reporter, so make sure that parking is not a problem.
- Does the facility have a room large enough to accommodate a news conference? Are there enough electrical outlets to accommodate cameras and audio plug-ins? Is a lectern available as well as chairs enough so that all reporters can be seated during the conference? You will be well served to secure the services of a professional audiovisual company to make certain that the electronic needs of the media can be met.

Room setup is important to the overall success of the news conference. Remember that you must accommodate broadcast and print reporters along with still photographers. Exhibit 7.1 offers a possible room arrangement for your consideration. Let's walk through this exhibit and explain its reasoning:

- The presentation area should be set up at the end of the room opposite to the entry door. You do not want pictures of a news conference showing people coming and going in the background.

EXHIBIT 7.1 Sample Room Setup

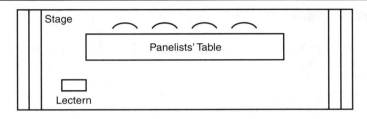

Leave this space open — 8–10 feet — so that heads are not blocking view.

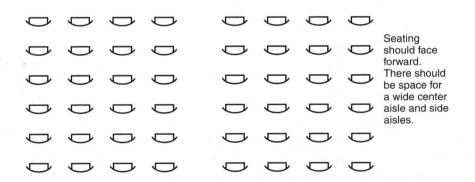

Seating should face forward. There should be space for a wide center aisle and side aisles.

This space will be used for camera equipment.

Greeting Table

(*Source:* Janine Reid Group, Inc.)

- At the top of the room, set up a riser or stage (if available) that is elevated a few feet above the ground. This elevation is important because photographers and TV camera crews need a clear view of company representatives. Place a table on the stage at which your company's conference participants can be seated. Also, place a lectern on the stage for them to use.
- Allow approximately 8 to 10 feet between the stage and the first rows of seating to ensure a clear view for all attendees. Also, leave a similar amount of space at the back of the room for TV cameras. Set up the seating in theater style with the chairs facing forward. Leave enough room for a wide center aisle and side aisles. Direct stragglers to seats along the side aisles.

Should you prepare handout material for the reporters?

Absolutely. Reporters are accustomed to receiving media kits at news conferences. A media kit is simply a folder that includes company information that the press is likely to request. A media kit should include copies of materials referred to during a news conference. If your organization does not have folders with the company logo on them, purchase plain folders from an office supply store, print your company's logo and name on labels, and affix the labels to the front of the folders.

Distribute media kits at a greeting table that you set up at the entrance to the room. Ask media representatives to sign in when they arrive at the conference. Keeping a list of all those who attended the news conference and the station/publication they represent will assist you when you send out updates about your crisis.

Most companies commonly include the following handouts in their media kits:

- Copies of all statements made at the conference.
- A list of the most frequently asked questions, accompanied by your answers. This list should help to increase the reporter's accuracy and also serves as a support sheet for your spokespeople.
- Fact sheet (see Exhibit 3.9). Don't forget: when you are preparing a fact sheet, try to think from the perspective of a reporter who is writing a story that involves your company. Include information that the press is likely to request. Keep it simple. Use short bulleted items rather than extensive paragraphs. Also, this is not the time for self-serving marketing text.
- Drawings or diagrams of the project.
- Stock and aerial photos, if available and appropriate.
- Brief biographies of company participants in the conference.

Who should participate in the news conference?

The first step in determining who in your company should be asked to participate in your news conference is to make a list of the messages you want to communicate and the questions that reporters will most likely ask. This process will drive the selection of your news conference's speakers. Let's investigate this process by returning to XYZ Construction's accident, determining what questions might be raised, and identifying the appropriate person to provide an answer:

1. Questions related to the injured workers and the accident that occurred last year should be answered by the president or another high-ranking company official.
2. Safety issues should be addressed by the safety director.
3. Questions related to the crane, rigging, scheduling, and neighborhood could be addressed by the project manager or superintendent.
4. Issues related to Mary Smith, the potential sexual harassment case, could be addressed by human resources.

The key player in any news conference is the corporate spokesperson—in XYZ's case, John Smith. As he is responsible for ensuring that the news conference runs smoothly, let's outline his responsibilities.

In a news conference setting, the corporate spokesperson serves as a panel moderator. Because John Smith's responsibility is to ensure the successful outcome of the news conference, he must establish the ground rules for conference panelists and the media at the outset. For the media, he must explain how the conference will proceed. Reporters need to know who is going to speak, how long that person will speak, and when the question and answer period will begin. A news conference should run no longer than 45 minutes. Most run from 30 to 45 minutes, including 15 to 20 minutes for the question and answer period. The standard procedure is that all panelists deliver their statements before the floor is opened to questions.

John Smith is also responsible for ensuring that the conference participants are aware of their roles during the conference. He has found that a convenient way to create this awareness is by preparing an agenda. If participants are unclear about their subject matter, he spells it out for them on the agenda. Also, he indicates the amount of time that each panelist has to discuss a particular matter, and he limits each one to this time—no more than 3 to 5 minutes. He urges everyone to be brief and to avoid jargon. When all preparations are complete, he opens the conference with the following statement:

> Good morning. Thank you for coming today. My name is John Smith, and I am [title] for XYZ Construction. We are here today to provide a briefing about the accident that occurred yesterday at the [project]. Because we know that you have a deadline, we will provide a few brief remarks before opening up the floor for questions. The conference will end at [30 to 45 minutes later]. At this time, I would like to introduce the people who will be providing you with an update of the incident. They are [title/name/topic to be covered].

The first person to be introduced should be XYZ's president. Companies fare better when top management responds immediately with a statement of genuine concern and commitment. When the media perceive a lack of concern from the company's leadership, this lack is reflected in the coverage of the event.

XYZ's president expresses the company's concern for the welfare of the injured employees and their families. In addition, the president expresses a commitment to finding out what happened and to making sure that it does not happen again and addresses any questions relative to the accident that occurred last year.

XYZ's safety director discusses the company's commitment to safety and the steps that it takes to ensure a safe work site. Additional comments can be made about the safety requirements the company places on its subcontractors, such as mandatory participation in all on-site safety meetings.

The project manager or superintendent addresses issues related to the selection of the mechanical subcontractor for this project (given the accident last year), as well as any questions related to the crane, such as where it was leased, who employs the operator, the date of the last inspection, and so forth. Additional comments can be made about the current status of the schedule.

The human resources representative from XYZ should have all of the facts relating to Mary Smith's conversations with the company with respect to possible sexual harassment charges and a detailed log of XYZ's actions in this area. The human resources representative must be prepared to deliver XYZ's position on sexual harassment in the workplace.

After all speakers have finished, John Smith opens the floor to questions. Maintaining the role of facilitator by asking reporters to identify themselves and the organization they represent, he then directs questions to the appropriate panelists. If necessary, he repeats each question before the appropriate party answers it. Professional courtesy grants reporters one or two questions before another reporter is recognized. If a particular reporter does not adhere to this courtesy, John Smith directs that reporter's attention to the other reporters.

At the designated time, he ends the conference by saying "Thank you for attending" and by announcing the company's plans for further communication.

As discussed in Chapter 5, your panelists may not be accustomed to speaking to the media, so it is critical to conduct a "stress rehearsal," if time permits. A news conference places a lot of pressure on company panelists, so it may be difficult for them to think clearly, a condition that could lead to unnecessary rambling. After you identify each panelist's message, break it down into three key points and coach him or her to return to them continually.

How do you notify the media of the news conference?

In a crisis, the media clamors for information; thus, you do not need to go to a great deal of trouble to invite them to the conference. Remember that reporters go in search of information if none is forthcoming, so you should hold your news conference within 24 hours of the incident. Typically, it is best to hold the news conference during late morning to allow reporters to meet deadlines. In a crisis, however, the time of day when you hold your conference is not a major consideration. If reporters are demanding answers, they will come whenever you hold your conference, within reason.

You have two ways to announce your intention to hold a news conference: (1) at the first statement *after* your buy-time statement or (2) through a media advisory, as illustrated in Exhibit 7.2. The sole purpose of a media advisory is to advise media outlets that you are holding an event that you would like them to attend. Provide a summary of the event, the reason it is being held, and details about its time and location. A media advisory is particularly useful if the vast majority of reporters are contacting you by phone. If so, send the advisory to media outlets via fax or e-mail, using the media list that you prepared as part of your crisis management plan. Media advisories should be printed on company letterhead.

If your company has not yet developed a media list, check with colleagues at other companies to determine whether they already have a list of general contacts that you can use. Other alternatives are to look in the local Yellow Pages and on the Internet under categories such as "broadcast television and radio stations" and "newspapers." Because you will not have time to identify the specific reporter assigned to cover your industry or story, send the media advisory to the city editor or newsroom

EXHIBIT 7.2 Sample Media Advisory

XYZ Construction Company
123 Jasmine Street
Anywhere, USA

FOR IMMEDIATE RELEASE   **CONTACT: John Smith**
[Date]   [Phone]

MEDIA ADVISORY

WHAT: XYZ Construction Company will hold a news conference to brief
                         media representatives regarding the accident that occurred at
                         [project] at 3:30 P.M. today.

WHEN/WHERE: [Hotel]

                         Conference Room [Name]

                         Address

                         [Date]

                         [Time]

SPECIAL INSTRUCTIONS: Free parking is available to the press in the building's
                         underground garage, located at the corner of 15th Street and
                         Blake. Be prepared to show your press identification. Take an
                         elevator to the 5th floor; the conference room is to your right.

# # #

(*Source:* Janine Reid Group, Inc.)

editor. The editor will assign the story to the appropriate person. Here are a few additional tips to ensure a successful news conference:

1. Start the conference at the appointed time. Promptness is important for journalists, who have deadlines and often have more than one story to cover. Remember to follow up with any journalists who request additional information by providing it to them when you said you would.

2. Videotape the news conference; for best results, secure the services of a video company. The tape will serve as an excellent training tool for your spokespeople in reviewing not only their performances but also their effectiveness based on the resulting news coverage.

Chapter Summary

- Use your best judgment when deciding whether or not a news conference is warranted. If you are not receiving numerous inquiries from the press, your crisis management team may be able to handle the situation without a conference. On the other hand, if you are deluged with requests for interviews, call a news conference.

- If possible, select a neutral site for your news conference and make certain that the facility can accommodate the electronic needs of the media.

- Invite the media by using a media advisory that lists the who, what, when, and where of the conference.

- Ask reporters to sign in so you have a record of their attendance. This list can also be used in sending them updates relative to your crisis.

- Prepare a media kit for reporters to peruse both during and after the news conference. Always include a list of frequently asked questions, accompanied by a company response to each.

- News conference panelists should be selected based on their level of involvement with the situation and their technical expertise.

- The first speaker should be the company's president (or another top official, if the president is not available), who emphasizes the company's concern about the situation and its commitment to its employees and the community. The corporate spokesperson should then take over the remainder of the press conference and act as the facilitator.

- Hold a "stress rehearsal" for anyone who will be a speaker at the news conference. Meeting the press is a highly stressful situation for most people; preparation enhances everyone's confidence.

8

COMMUNICATING WITH YOUR VARIOUS AUDIENCES WHEN THE NEWS IS BAD

Who are your audiences and why do you need to communicate with them?

As discussed in Chapter 1, an audience is defined as anyone who can have an effect on your business or reputation. When a crisis strikes, insurance companies may cover the immediate damage, but their coverage does not include reclaiming your reputation and credibility in a competitive marketplace. This effort is your responsibility; it should be taken seriously and addressed quickly at the outset of a crisis. If you do not take the time to communicate your side of the story and to explain what you are doing about it to your various audiences, they will have no choice but to believe what they read and hear from the news media and from your competitors. Conversely, open and honest communications, executed quickly and consistently throughout your crisis, can maintain and perhaps enhance your reputation and credibility.

Take a moment to review Exhibit 8.1 and determine which audiences could have an effect on your business or reputation in the event of a crisis. Next, customize this list for your company by adding audiences or deleting ones that do not apply. A great resource to assist you in this process is your marketing contact list—sometimes referred to as a marketing information database. If you do not have such a list, now would be a good time to develop one.

Once you have identified key companies within each audience, you should then hold someone within your company responsible for making contact with the key influencer from each company during a crisis. The ideal match is someone who has an existing relationship with the influencer.

EXHIBIT 8.1 Company Audiences

Board of directors/shareholders

Employees

Clients (current, potential, past)

Other contractors, architects, engineers

Unions

Suppliers

Insurance company

Surety company/banks/other lenders

Stock analysts

Government regulators

City/county officials

Site neighbors

Opinion leaders

Action groups

Others . . .

(*Source:* Janine Reid Group, Inc.)

What is your message and what is the method of delivery?

Now that you have defined your audiences and matched each key influencer with a company representative, the next step is to determine the message that will be delivered. The process gets interesting here because your message may vary slightly by audience, depending on the effect the crisis has on each. To illustrate this process best, let's return to XYZ Construction's crisis and walk through this logic step by step.

Let's assume that it is 9:00 a.m. on the second day of the crisis. John Smith, the company spokesperson, has just delivered his statement outlining what few facts he was able to gather and verify. He has answered questions from the reporters and felt that he did a reasonable job of communicating compassion, concern, and a dogged determination to find the cause of the accident. Let us also assume that the resultant press coverage pointed the finger of guilt at XYZ for endangering the lives of its workers by hiring a mechanical subcontractor with a questionable safety record. Based on this assumption, John refers to Exhibit 8.1 and determines which audiences could be affected by the negative media attention. Those audiences might include the following groups:

Board of directors/shareholders

Employees

Clients (current, potential, past)

Architects, engineers, and other contractors

Suppliers

Insurance company

Surety company/banks/other lenders
Site neighbors

John knows that XYZ's insurance company, the project owner, and the applicable unions were notified immediately after the accident occurred, so they are not included on this list.

Now, let's take each one of these audiences individually and discuss the following points:

1. Why was this audience selected?
2. What is the message for this audience?
3. Who should be the XYZ messenger?
4. How should the message be delivered?

Board of directors and shareholders

Why Was This Audience Selected? The board of directors and shareholders of XYZ, both internal and external to the company, must be notified immediately of the crisis and of potential negative effects that could affect XYZ's bottom line, reputation, and credibility in the marketplace. The audience should be kept in the loop of communication because they will most likely be queried by other groups or individuals, and their response should be consistent with the company message(s).

What Is the Message for This Audience? The board of directors and shareholders should receive all of the facts that are known at this time and be informed of all of the potential ramifications of the crisis.

Who Should Be the XYZ Messenger? The president of the company or another top official should deliver XYZ's message.

How Should the Message Be Delivered? Every effort must be made to initiate this contact as soon as possible via telephone or personal visit.

Employees

Why Was This Audience Selected? The employees of XYZ are a primary communication vehicle to the outside world. If XYZ is open, honest, and committed to frequently updating employees about the crisis, they will act as loyal ambassadors who support their company's actions. However, if XYZ elects to withhold information or skirt issues, it may find that employees become highly critical and can contaminate the opinions of other audiences.

What Is the Message for This Audience? XYZ's message should cover the following information:

1. The facts, as you know them, and the current status of the investigation.
2. Concern and compassion for those injured in the accident.

3. Assurance that the health and safety of all employees is XYZ's top priority and that everything is being done to determine the cause of the accident.
4. The name and telephone number of XYZ's spokesperson in the event an employee is approached by a reporter.
5. The name and telephone number of an XYZ crisis management team member who can be contacted with questions related to the incident.
6. Notification of when the next update will occur.

Who Should Be the Messenger? The president of XYZ, or another XYZ top official, should originate all communications to XYZ employees.

How Should the Message Be Delivered? The message should be delivered via the fastest vehicle possible to all XYZ offices and job sites. This delivery could include a special meeting, broadcast voice mail, broadcast e-mail, fax, notice on bulletin boards, or any other method you can identify. Exhibit 8.2 is a sample memo that XYZ could use.

Clients (current, potential, past)

Why Was This Audience Selected? A client's reputation is at stake during a crisis, just as a contractor's reputation is in danger. XYZ's clients will closely scrutinize the management abilities of the company throughout this crisis, and they will be among

EXHIBIT 8.2 Memo/Fax/E-Mail to XYZ Employees

MEMO

TO: All XYZ Employees
FROM: Hugh Grant
DATE: [Date]
RE: Accident at the [project]

As you may be aware, we experienced an unfortunate accident yesterday that occurred during the lift of a cooling tower to the eighth floor. We regret to report that three people were injured.

Larry Lynn, a carpenter apprentice for XYZ Construction, suffered head injuries and is in critical condition. Ron King, who is a laborer for ABC Mechanical Company, has internal injuries and is in serious but stable condition. Both Larry and Ron are at University Hospital. Juan Lopez, also a laborer for ABC, was released from the hospital last night with a broken leg. We are staying in constant contact with the hospital and will keep you updated on Larry's and Ron's conditions.

We have all been touched by this accident and, at this writing, we do not know the cause of the failure. We are working closely with the authorities as they proceed with their investigation. In addition, we are conducting an independent investigation to help determine exactly what happened so we can make certain that it never happens again.

As you know, this accident has drawn some attention from the news media. If you are approached by reporters, simply refer them to our spokesperson, John Smith, at [phone]. In the meantime, I will personally keep you informed of any future developments.

(*Source:* Janine Reid Group, Inc.)

its severest critics. XYZ has a narrow window of opportunity to turn any potential adversary into an ally, and that opportunity is based on timing and message content.

Let's take a look at the three types of clients defined above and determine the impact of XYZ's crisis on each.

Current Clients. When a crisis strikes, many of XYZ's current clients will ask themselves a lot of questions:

- Could this accident happen on our job?
- Is the same mechanical subcontractor working on our job?
- Will the media knock on our door to inquire about why we hired XYZ?
- If XYZ is guilty, will it still be in business to complete our job?
- Can we continue to trust XYZ's judgment and decision-making capabilities?

To maintain or restore a strong level of confidence, XYZ must be able to anticipate and respond quickly to all such questions that this audience may present.

Potential Clients. This audience includes companies that have received or will receive a proposal for services from XYZ. Nothing can scare a client away faster than being solicited by a company that is receiving negative media attention. Again, a client's reputation is also at stake when a crisis occurs, and a potential client will probably think twice before hiring a company in the midst of a negative situation.

Past Clients. This audience is an excellent resource for future business as well as a reference and testimonial provider. This audience must be able to maintain its level of confidence in XYZ.

What Is the Message for This Audience? Owners and developers will expect XYZ to have the ability to manage any situation; therefore, XYZ must communicate that it is clearly in control of the situation and is doing everything in its power to determine the cause of the accident.

Who Should Be the XYZ Messenger? Ideally, the messenger should be the XYZ individual who has the strongest relationship with each particular client. This person could be the director of business development, the project manager or superintendent, the vice president of operations, or the president.

How Should the Message Be Delivered? A telephone call or personal visit should be the method of delivery for all current and potential clients. A letter, fax, or e-mail could be used for past clients. Make certain that all written communication is done on a personal level. Do not use the salutation "Dear Client" or "To Whom It May Concern." Those salutations smack of mass distribution and could have a negative impact—all the more reason to make certain that your database is up to date and that you have the ability to personalize letters or memos. Exhibit 8.3 outlines a sample letter to send to XYZ's various audiences.

EXHIBIT 8.3 Sample Letter to XYZ's Audiences

XYZ Construction Company
123 Jasmine Street
Anywhere, USA

[Date]

Jack Green, President
Big Owner Corp.
630 Lafayette Avenue
Anywhere, USA

Dear Jack:

As you may be aware, we experienced an unfortunate accident on [date] at our [project]. The accident occurred when a crane was lifting a cooling tower to the eighth floor. We are sorry to report that three people were injured in the accident. Two of the workers are expected to make a full recovery; however, the third remains in critical condition.

As you can well imagine, we are all deeply touched by this accident and are diligently working with the authorities to determine the cause. We remain steadfast in our commitment to safety, and we will do whatever it takes to make sure such an accident never happens again.

We will keep you informed of our progress as the investigation continues. In the meantime, please contact me with any questions or concerns.

Sincerely,

Hugh Grant
President

HG/mp

(*Source:* Janine Reid Group, Inc.)

One last word on communicating with potential clients. Let's fast-forward one year and imagine that XYZ is at the presentation table with an owner or developer and is trying to secure a project. XYZ should not, for one minute, think that its prospective client may not have heard of the accident or has an incredibly short memory. XYZ must be fully prepared to discuss the accident and what it learned from the crisis; it must give assurances that, to the best of the company's ability, a similar situation would never occur. XYZ should not try to hide from the issue because, frankly, hiding anything from anyone in today's competitive world is virtually impossible.

Architects, engineers, other contractors, and suppliers

Why Were These Audiences Selected? Architects and engineers can be an important source of business and referrals for XYZ. These two audiences can influence the buying decision of an owner or developer in some cases, so does a subcontractor. The industry rumor mill can run wild during a crisis, so the most effective way to get the XYZ story out is to communicate consistently with the architects, engineers, subcontractors, and suppliers that have relationships with XYZ.

What Is the Message for These Audiences? The message content should reinforce the message XYZ's client base.

Who Should Be the XYZ Messenger? As with XYZ's client base, the messenger should be the XYZ individual who has the strongest relationship with the key influencer at each of these audiences.

How Should the Message Be Delivered? A telephone call is recommended for those firms that currently work with XYZ or that have the potential to work with it in the immediate future. All others can receive a letter, memo, fax, or e-mail.

Surety company/banks/other lenders

Why Were These Audiences Selected? Any financial partner of XYZ should be notified as quickly as possible and kept informed of the investigation as it proceeds.

What Is the Message for These Audiences? As with the board of directors and shareholders, XYZ's financial partners should receive the straight facts that are known at the time and the negatives that could result from the crisis.

Who Should Be the XYZ Messenger? Once again, the president or another top official of the company should deliver the message.

How Should the Message Be Delivered? The communication should take place as quickly as possible via telephone.

Site neighbors

Why Was This Audience Selected? If you recall, in our hypothetical situation, XYZ's project in crisis is located near a residential area, and the residents have complained about the noise, dust, and truck traffic. This crisis gives the neighborhood group leader a reason to complain even more loudly and take his complaint to a reporter or anyone else willing to listen.

What Is the Message for This Audience? XYZ should make certain that the neighborhood group leader is kept advised of inconveniences that might occur as a result of the accident. This might include street closures and additional noise resulting from the cleanup procedure. The site neighbors must be assured that their safety and health will not be jeopardized in the cleanup process and that XYZ will maintain consistent communications with the neighborhood group leader throughout this process and beyond.

Who Should Be the XYZ Messenger? The best candidate is the individual who has established a relationship with the neighborhood group leader. This person is most likely XYZ's project manager or superintendent.

How Should the Message Be Delivered? The communication should take place via a telephone call. However, the neighborhood group leader may show up on site, demanding answers to a multitude of concerns. If so, a member of the crisis management team should make the time to meet personally with this individual.

Now you may be thinking, "I don't have time to make all of these contacts because my world is coming down all around me!" Yes, that is true, but as we previously discussed, if you do not communicate your side of the story quickly, all of these audiences will believe what they read and hear from the news media. Therefore, it is up to your crisis management team to determine which audiences should be contacted and when that communication should occur. Exhibit 8.4 provides a model action plan to help you organize this effort.

How often should you update your audiences on the status of your crisis?

A good guideline is to provide an update every time you have new information to release to the news media. That way, everyone receives the same message at the same time.

At some point, believe it or not, the media attention will begin to subside. Please do not believe for one instant that your various audiences are going to forget about your crisis just because it is not being reported in the papers or on local news stations. You must always be prepared to address any questions that may arise regarding your crisis. Also, make certain that your employees are always kept in the information loop about the status of the injured and the accident investigation, because your employees are your communication vehicle to the outside world.

Is print advertising a good medium to use to communicate your message during and after a crisis?

The answer to this question depends on what you are trying to accomplish. Let's take a look at advantages and disadvantages of using advertising as a communications medium. First, the downside:

- Advertising is impersonal in nature and caters to the general audience of the publication, not necessarily the specific audience you may need to reach.
- Advertising delivers a shotgun approach, that will broadcast your crisis to all who receive the publication—whether they are aware of your situation or not.
- Advertising is so dominant in publications today that unless you spend a lot of money on ad development and size, your message will most likely be missed. Also, advertising lacks credibility because it is paid for and because the advertisers can write the copy exactly they way they wish; therefore, an ad may be perceived as propaganda, not fact.

EXHIBIT 8.4 XYZ Communications Action Plan

Effective 9:00 A.M. on the second day of the crisis

Audience	Personal Contact	Phone Call	Letter	Fax/E-Mail	Flyers or Advertisement	XYZ Contact	Timing
Board of directors shareholders	X	X				President/ Top Official	Immediately
Employees	X	X		X		President/ Top Official	Immediately
Current and potential clients	X	X				Assign	Today
Past clients			X	X		Assign	By tomorrow
Current and potential architects and engineers	X	X				Assign	Today
Past architects and engineers			X	X		Assign	By tomorrow
Other contractors/suppliers			X	X		Assign	By tomorrow
Surety company/banks/ other lenders		X				President/ Top Official	Today
Site neighbors		X			X	Project Manager/ Superintendent	Today

(*Source*: Janine Reid Group, Inc.)

125

Now, let's take a look of some of the advantages of advertising:

- If your company was the victim in a crisis—from, say, a natural disaster, sabotage, or workplace violence—an ad could be a useful vehicle for acknowledging the individuals and companies that provided assistance during the crisis.
- If your crisis creates an inconvenience to surrounding areas and you wish to reach a broad audience with your message, advertising can be a very effective vehicle. However, as discussed earlier, not everyone reads advertising, so you may wish to back up this effort with flyers or door-hangers to ensure that your message reaches your target audience.

If we return to Exhibit 8.4 and walk through the communications action plan for XYZ, you will notice that the distribution of flyers or the use of advertising is the recommended additional communications vehicle for the site neighbors surrounding the project. As we discussed, this audience needs continual reassurance that the health and safety of their neighborhood is not being compromised as a result of XYZ's crisis.

On balance, I have experienced a much higher success rate using personal attention to a company's audiences during a crisis than I have using advertising.

Should you contact your audiences if your crisis has not gone public?

There is no need to create external concern if the issue can be contained internally. However, news does have a way of leaking, so it is a good idea to be prepared for information to wander outside your doors into the ears of your audiences. Should this occur, be prepared to plug the leak as quickly as possible through direct and consistent communications with the interested parties.

Chapter Summary

- Consistent communications with your various audiences during a crisis will deliver your company positive results. If you fail to implement such a communications program, your audiences will believe everything they hear and read from both the news media and the rumor mill.
- Develop a marketing information database that lists all contacts within your respective audiences. This database will be used frequently during your crisis and must be current with respect to contacts, addresses, and phone numbers.
- Always personalize written communications from your company to your various audiences. Generic salutations are perceived as impersonal and invite a cold reception. Each letter should communicate an invitation to contact the sender with questions about the crisis.

- A crisis can have a long shelf life, so a company must always be prepared to discuss, with any of its audiences, the situation and the lessons learned from it.
- The use of advertising as a communications medium during and after a crisis can be effective in some circumstances and a cash drain in others. Weigh this option carefully and base your decision on the target audiences you want to reach and the message you wish to communicate.

9

DEVELOPING POSITIVE RELATIONS WITH THE NEWS MEDIA THROUGH A PUBLIC RELATIONS PROGRAM

Public relations is only one component in a company's overall marketing program. Unfortunately, this chapter cannot give this discipline the justice it deserves. With so many advantages to a strong public relations effort in the design and construction industry, an entire book could be written on the subject. This chapter, however, focuses on why a public relations program is important to a company prior to a crisis.

In previous chapters, we talked about a reporter's ability to dig for information on your company when a crisis strikes. One way reporters do this research is by searching computer databases, such as Lexis-Nexis. These information banks are capable of retrieving stories that were printed and, in some cases, broadcast from media outlets ranging from local and national newspapers to association newsletters and trade journals. Thus, you can be certain that reporters can unearth anything that the media has had to say about your company.

This chapter's goal is to talk about how an ongoing public relations effort allows your company to deposit positive information into this "bank" to reflect the good things you do and the contributions your company makes to the community. Please understand that it does not matter if the articles are simply announcements about your staff or a commemoration of an anniversary; this kind of information begins to put a face on your organization by providing the reporter with a key to your company's values. If, on the other hand, the only background reporters can find about your company is negative, they cannot receive a good impression of your company.

Companies that pursue an ongoing public-relations program with their local media prior to a crisis typically have a credibility edge over their less savvy competitors, should a crisis occur. However, a public relations program does not guarantee immunity to negative coverage because you have no assurance that the reporter with whom you developed a relationship is the one who covers your crisis. Also, if you mishandle a crisis, no amount of public relations can erase the immediate damage. As emphasized before, the best way to protect your interests is to act in a proactive and ethical manner.

What does *public relations* mean?

In the design and construction industry, public relations is typically viewed as soft and fuzzy and is, therefore, defined in one word—overhead! There, I said it. That ugly eight-letter word really means "If you can't show me results today, I'm not interested." Now, before attempting to support or refute this definition, I would like to talk about the true definition of public relations and its oh-so brief history in the design and construction industry.

Practiced ethically, public relations has the ability to influence the attitudes of a company's audiences. Therefore, public relations is nothing more than strategic communications that support your marketing goals.

Once, in a communications class, I had a professor who said that the secret to a good public relations program is to "do good things and get credit for them." I never fully understood the power in that statement until I went to work for a general contractor who did good things but never took credit for them. This company was heavily involved with the community and donated untold hours to causes and outreach programs. It extended financial support as well as employees' time to a multitude of nonprofit organizations throughout the state where it operated. The company had a high degree of integrity and a reputation for quality. It hired the best people in the industry and invested in research and development to provide its clients with the best possible product, not necessarily at the lowest cost.

The owner of the company was a low-profile person and enjoyed doing good things for people and organizations, but he was not a showboat and did not want to advertise his good deeds. So the company did all of these great things but never told anyone about them. Now, this particular company enjoyed a great deal of success during the days when business was done on a handshake, but the market experienced a downturn and the company was sold. The moral of this story is that today's competitive marketplace dictates a more aggressive approach in getting your name out in a positive fashion.

As discussed in previous chapters, a poorly handled crisis can cause serious damage to a company's reputation. However, positive media attention can generate business and goodwill. So the sword clearly has two edges.

Why don't more companies use public relations programs?

Unfortunately, the vast majority of companies involved in the industry do not promote their efforts through the media. Why? The answer is that public relations can be a frustrating journey for the following reasons:

- It takes time and you do not see immediate results. Well, there you have it. To those involved in the design and construction industry, time is money. If the question "What has it done for me today?" cannot be answered by showing a contribution to the bottom line, the program will have a very short life.

- If you hire a public relations agency to promote your company, you may become frustrated because of the agency's lack of industry knowledge. This ignorance creates a time burden for you because someone in your company must educate and monitor the agency's every move. So you spend a lot of time ramping the agency up, waiting for minimal results, then firing the agency and saying, "I knew this wouldn't work!"
- Someone in your company may take on the challenge of working with a public relations agency but become frustrated because everything submitted to the media either gets ignored or diluted to the point of being unrecognizable.

Are these valid reasons? Of course they are; however, they do not give the industry an excuse to dismiss this valuable discipline for two reasons:

1. Those in the design and construction industry must start taking credit for all of the good things they do.
2. This action will have an overall positive effect on the industry's image—which is sorely in need of improvement!

An organized and consistently applied public relations program is clearly worth the time and effort because of the third-party endorsement that the media offers. Every company involved in the industry has a story to tell and, frequently, good things to say, so let's take a look at how that disseminating information can work to the industry's advantage.

First, we review ideas on an individual company level. Then we talk about the effect that those efforts can have on the industry in general.

How do you take credit for the good things you do?

The first step is to establish a public relations program. Such programs come in all sizes and shapes. Your company's efforts can range from the simple to the complex, depending on your resources. Some companies elect to hire freelance people or agencies to develop and implement their programs, while other companies may assign these duties to in-house employees. No matter which route you select, certain steps need to be taken to get you to your destination.

Develop your message

The old saying "If you don't know where you are going, you aren't going to get there" applies here. As you are probably aware, public relations is only one component in the marketing mix of a company; therefore, the marketing plan drives the public relations program. Thus, the program must focus on marketing goals. For example, you may wish to pursue the following goals:

1. Create a new image. Perhaps your company has just opened its doors and you would like to announce this event to your various audiences. Or suppose you have opened a district office in a geographic area that knows nothing about your company. Both of these events offer great opportunity—first, to communicate with your audiences by using a third-party endorsement and, second, to create a relationship with the local media.

2. Enhance an existing image. This effort adds to a maintenance program that a company may have already established with the media. The idea is be on the lookout for positive stories about the company to get the attention of reporters. These stories might focus on a new technology the company uses or an employee who has made a significant contribution to the company, industry, or community. Opportunities for such stories are everywhere. All you need is awareness.

3. Change an existing image. Let's say that you have a reputation for designing or building churches in your locale; however, your marketing study reveals that healthcare facilities promise more contracts and will continue to do so for the foreseeable future. A public relations effort can help you change your image from a company with church experience to one that also has a working knowledge of designing and building healthcare facilities. So what if your company has never designed or built such facilities? You will probably hire people with the necessary experience and promote them as your experts.

4. Educate your audience(s). Nothing sells better than brains! Publications like to receive educational and how-to articles. Always look for ways to differentiate yourself by educating your audiences.

Companies receiving positive media coverage have made an effort to inform the media of milestones that reflect well on their expertise and employees. Review the following list of opportunities and think about how many times you have seen similar stories covered by the press.

- Contract awards, groundbreakings
- Topping out, project completions
- Grand openings, dedications
- Company anniversaries
- Feature stories, human interest stories
- Entries into new markets
- New service offerings
- Speeches or presentations on topics of local/national interest
- Industry awards
- Personnel promotions, new hires

- Sponsorship of community activities
- Special areas of expertise
- Notable employee achievements
- Corporate expansion
- Subsidiary formation
- Innovative approaches
- Paper, article, or book publication
- National association affiliation

The list does not stop here. Opportunities to tell your story abound everywhere. Contact your local and national associations to get their assistance in developing and placing stories. Many of these associations are proactive in the area of media relations and can offer valuable insight about the types of stories, articles, and press releases that certain media seek. Also, because many reporters contact these associations to find sources for their positive stories, you could position your company as the local expert in one or more areas.

Identify the audiences that you are interested in reaching

Once you develop your message, the next step is to determine the audience(s) you are most interested in reaching. If, for instance, you are interested in marketing your services to hospitals, then you want to target publications that cater to the healthcare industry. Now, let's take a timeout here. Before you submit any information to a publication, make certain that you fully understand the publication's content and focus by reading several issues cover to cover. Call the editor and ask for a list of editorial guidelines. These guidelines will educate you about submission requirements so you can make your material more acceptable. You may find that some publications are 100% staff written and that outside editorials are not accepted. If so, you can pitch your story idea to the editor. If it meets the editor's requirements and is felt to be newsworthy, a reporter will be assigned to write the story. Most publications have sections that are dedicated to industry news, such as releases on company formations, hiring and promotion announcements for mid- to upper-level management, corporate expansions, and so forth.

Many advantages develop from an ongoing public relations campaign, and they go beyond merely making deposits in the database bank:

1. Compared with the medium of advertising, public relations is relatively inexpensive and clearly packs more power.
2. Positive recognition in the news media affords the company third-party endorsements, which deliver credibility and prestige.
3. Reprints of positive attention can be used for direct-mail purposes as well as inclusion in proposals. Again, you achieve third-party endorsements that differentiate you from your competitors. Do not forget to send your releases to the

industry press as well as the general press. You may find that trade journals are more likely to pick up your releases, and the reprint value is still there.

4. The positive messages communicated through your public relations program also have positive effects on the overall industry.

How can your public relations effort have a positive effect on the industry's image?

Perhaps a better a way to ask that questions is "Does the industry's image need help from everyone who participates in it?" The answer is an unequivocal *yes!*

Construction is a highly visible industry. Everything it does, good or bad, is out there for all to see. When something negative happens, whether a fatal job-site accident or seemingly endless road repairs, the industry as a whole takes the blame. However, the reverse seems to occur when the industry does something good. Remember the contractors who dropped everything they were doing and headed for the Alfred P. Murrah Federal Building in Oklahoma City to lend assistance and the hundreds of contractors who donated time and materials to clean up and repair Columbine High School in Littleton, Colorado, after the massacre? Unfortunately, those deeds garner little attention, yet they occur frequently all over the world. Now I am not suggesting that we capitalize on someone's misfortune, but we must learn to look at each event with fresh eyes to determine how we can get credit for the good things we do.

To illustrate this point, the editorial page of *ENR* ("Colorado's Constructors," 1999) applauds the 130 firms that donated about a half million dollars in labor, materials, and equipment to rehabilitate Columbine High School. As many of you might remember, two students armed with bombs and guns killed 12 students and a teacher and injured 23 others on April 20, 1999. The closing paragraph of this editorial says it all.

> Unquestionably, this was a remarkable effort from this industry that seldom toots its own horn, seeks reward for humanitarian efforts or curries up to the general media to put a "positive spin" on what it does. Probably the best reward the construction industry could receive would be for a few of the Columbine's 1,900 students to decide that an industry whose people gave so freely of their own time and energy to reconstruct a physically scarred school is an industry in which they'd like to work.

The design and construction industry has typically been its own worst public relations agent. The time for change is past due. Now we must broadcast the message that the industry offers something for everyone—from working in the field using ideas and hands to owning a company. Not only is positive media coverage good for business but also it has a positive effect on the entire industry.

So the real question becomes "Is a public relations program a leap of faith?" In most cases, the answer is *yes* because you will not realize the instant gratification that the industry demands. However, as we discussed in previous chapters, the media hold the cards of power and can hurt you during a crisis, but they also have the ability to enhance your credibility and reputation by reporting on the good things that you do. Remember, some day you will need to make a withdrawal from the "information banks," so your best interests are served by continually making deposits. The benefits are worth the effort—for your company as well as for the future of the industry.

Chapter Summary

- Reporters have the ability to access computer databases to discover positive and negative information about your past. An ongoing public relations effort gives you a credibility edge with the media by filling those databases with good things about your company and the contributions it makes to the community. This edge is vital in a crisis.
- Public relations is an underused component of companies' marketing plans because the results of such efforts are subtle. However, they are also powerful because they secure third-party endorsements and influence a company's audiences positively.
- The secret to good public relations is to do good things and take credit for them.
- As a supporting player in the marketing mix, public relations has the ability to create a new image, enhance an existing image, change an existing image, and educate your audiences.
- Tap your local and national associations for input and assistance in developing and placing stories. Your associations can also help position your company with the media as an expert in certain areas of the design and construction industry.
- The positive messages communicated through your individual public relations effort have a positive effect on the image of the entire industry.

10

TRAINING YOUR EMPLOYEES TO BE PREPARED FOR A CRISIS

Why does a company need to train its employees in managing a crisis?

Throughout this book, I stress the importance of having a written protocol and an established crisis management team. However, the reality of the construction industry has shown that, in the throes of a crisis, many people do not refer to a plan or remember who to call for assistance. When the stakes are real, a company that has trained its employees in the art of managing a crisis will reap the rewards of the sports principle that says "You will play the way you have practiced." Trained employees use the common sense and instinct that were honed during their last training session, whereas their untrained counterparts most likely fall victim to the intense stress and demands that a crisis creates.

Who should participate in a crisis management training session?

Attendance should be required of the core crisis management team and as many employees from each project crisis management team as possible. Additionally, all salaried personnel should participate in the training session because they will be the support group for the entire team. The higher your attendance in a training session, the stronger the support you will receive for your planning effort. Even though closing your doors or shutting down projects for a training session is impossible, you can hold repeat sessions in an effort to train as many people as possible.

You can also include your insurance company, broker, public relations counsel, attorney, and any other outside entity that you think would benefit from the training.

What training method should be used?

The secret to delivering a successful training session is to keep it highly interactive, informative, interesting, and as brief as possible while allowing enough time to cover the material sufficiently. People have short attention spans when it comes to lecture-based training that espouses theory and examines detailed case studies. To keep trainees' attention, offer and ask for participation throughout the entire session. Participatory training increases attention and retention—two critical components to effective training. Exhibit 10.1 reinforces this theory by showing a model of passive and active learning. As you will note, the highest percentage of retention occurs at the bottom third of the triangle, where group interaction is introduced.

What material should be covered?

The following sample workshop agenda is presented in an effort to achieve the highest retention rate possible. This workshop incorporates the entire content of Exhibit 10.1; however, the majority of the session embraces the lower third of the model, where the highest retention rate is realized. The only component that is not addressed is the very last one—"Doing the real thing." With luck, that component will never be required.

Sample workshop agenda

The "Crisis Audit" Exercise.   Let's say you are the instructor. The first point participants need to understand is what constitutes a crisis and what one might look like in their own company. Thus, review with them the definition in Chapter 1: "A crisis is any incident that can focus negative attention on a company and have an adverse effect on its overall financial condition, its relationships with its audiences, or its reputation within the marketplace." You can elaborate on this definition by customizing it to your company.

Next, the participants are broken into groups based on project assignment, department, and so forth. Each group is asked to do a crisis audit and create a list of ten potential crises that could happen at 9:00 a.m. tomorrow on their project, within their department, or to the company in general. Participants can be provided with a list of potential crises to select from, as outlined in Chapter 1, or they can create a list of their own.

Once the members of each group have developed their Top Ten list, they must reach a consensus on a priority ranking. In other words, this list is rearranged to present the most likely crisis, which is numbered *one,* to the least likely crisis, which is numbered *ten.*

Next, each group appoints a spokesperson to present the group's findings to all of the participants, explain why each crisis was selected, and explain the ranking of each.

The goal of this exercise to create an awareness of the company's daily risks. Risks can be avoided if they can be identified, but we need to teach the process of identification so that prevention can follow.

EXHIBIT 10.1 Cone of Experience

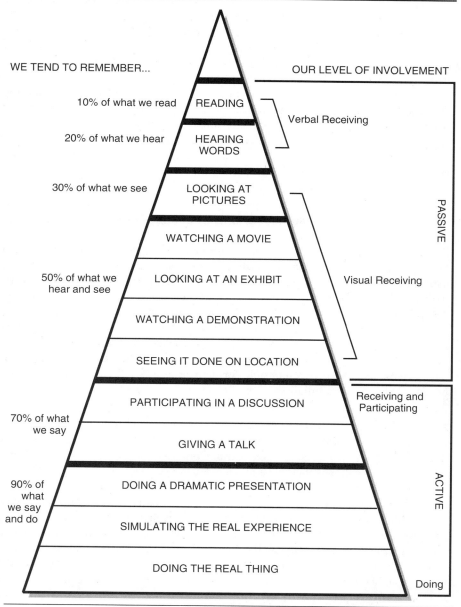

(*Source:* Developed and revised by Bruce Nyland from material by Dale Edgar)

The "What-if" Analysis Exercise.   The "what-if" analysis exercise requires each group's members to focus on the number-one crisis that was selected in their crisis audit. The assumption is made that the crisis just happened and that they must quickly determine, as a group, what chain of events could occur as a result. This assessment is accomplished by constructing a "what-if" analysis, as shown in Exhibit 1.2. As you will note in this exhibit, a structural collapse is indeed a crisis. In addition, it can create a domino effect that can quickly get out of control.

The participants in each group should take their number-one crisis and develop their own "what-if" analysis by using any written form they wish—perhaps a wishbone diagram, a branched-tree diagram, or simply an outline. Then, to the class, they should present their findings and the rationale behind their decisions.

The goal of this exercise is to teach the participants how to anticipate spin-off crises before they occur so they can be eliminated or minimized. This exercise develops critical thinking skills as well as an ability to anticipate the unknown.

Introduction of the Team Leader and Spokesperson.   Now is the time to introduce the responsibilities of the team leader and spokesperson. Do so before you present the simulation exercise, because all of the participants must understand that the team leader is the internal organizer in a crisis and that the spokesperson is the external communicator. Chapter 2 provides a full description of the team leader's and the spokesperson's responsibilities. By the conclusion of this part of the training, everyone should have a clear understanding that the team leader must be contacted at the outset of a crisis or of any threat of a situation becoming a crisis. Also, trainees must understand that the spokesperson is the only person authorized to speak to the news media on behalf of the company during a crisis, with the exception of a temporary spokesperson who is responsible for delivering a buy-time statement. Again, refer to Chapter 2 for a full discussion of this topic.

A Tabletop Simulation Exercise.   Simulation-based training embodies the theory illustrated in Exhibit 10.1—we learn by doing. Although not everyone learns this way, people enjoy an accelerated learning process that invites them to jump into a situation and challenge their existing knowledge.

A tabletop simulation is a facilitated discussion of a hypothetical crisis scenario and of the response strategies necessary to bring the situation under control as quickly as possible. Should a real crisis subsequently occur, participants' decision-making skills will be sharper and they will have learned to anticipate events rather than acquiesce to a series of surprises; this will help reduce mistakes. The participants will also learn how to use the crisis management plan as a tool to guide them through the initial stages of a crisis to its conclusion. The introduction of the plan is discussed in the next section.

How do you develop a tabletop simulation exercise?

Here are guidelines for creating a customized simulation exercise.

Developing a Scenario Idea.   The most important part of developing a simulation is to use a scenario that participants will find realistic. This quality heightens the level of involvement and enhances the learning process. If you are having a hard time coming up with a potential crisis, review the list of crises in Chapter 1 to get your imagination percolating.

Building a Story.   Create a scenario around a hypothetical company that is similar to the size and scope of work of your company. Give a brief background of this company, including the location of its home office, any district or regional offices, and the type of work it performs. Next, describe a project or incident that is currently underway, then create your crisis around it. Make the crisis the last item in your scenario and present it as something that has just occurred. The crisis should resemble an issue that could be a clear threat to the company's reputation and financial well-being. The example of XYZ Construction, as shown in Chapter 5, illustrates this process.

As you build your scenario, present just enough information to give the participants a fair grasp of your hypothetical company, but avoid going into too much detail. You want the information to be a bit vague and sketchy because in the early stages of a crisis, information is almost always incomplete or faulty. Be sure to include realistic challenges, such as notification of a family in the event of an employee fatality or an environmental spill that requires the evacuation of an entire neighborhood. Also, be sure to add the dimension of having a reporter show up on site to cover the crisis so you can see how the participants respond to pressure from the news media. You can include additional challenges, such as staging the incident at night, over a weekend, on a holiday, or at a time when a key crisis management team member is out of the country.

Involving the Participants.   The simulation is an active learning experience and, therefore, requires participation from everyone. In order to achieve this involvement, you need to decide which members of your crisis management team might become involved in your hypothetical crisis. For example, if you refer to the XYZ Construction crisis outlined in Chapter 5, you may elect to have six people on your team, including a team leader, spokesperson, project manager, superintendent, reporter, and perhaps a neighborhood group leader. These team members are called "players."

Once you have selected the players in your simulation, you need to develop "player cards," which outline their backgrounds with your hypothetical company and their involvement with the crisis. Exhibit 10.2 illustrates a sample player card for the project manager. As you can see, the participant who draws this player card will have a clear understanding of his or her background in the scenario being played out. The card also states what the project manager was doing at the time of the crisis. Use this idea to generate other cards for the remainder of the players you have selected for your simulation.

Conducting the Simulation.   Now, let's proceed to the actual implementation of the simulation. Your training room should be set up either in a classroom style or with round tables. Each table should be set for the exact number of players required

EXHIBIT 10.2 Sample Player Card

YOU ARE THE PROJECT MANAGER

You have worked for XYZ Construction for 11 years and have come up through the ranks the hard way. The president of the company has told you that you are the best project manager this company has ever seen. That is a great compliment, but with the title comes responsibility. This project is only one of the four other jobs you oversee—but it is the toughest. The owner has made some heavy demands on your time, and you are doing your best to keep him happy.

You are on site today to talk with the neighborhood group leader, who has been a thorn in your side since dirt was first scratched. No matter what you do, the complaints never stop. First it's the noise, then the dust, then the traffic. You are at the end of your rope, trying to figure out how to make everyone happy, and you have not received much support from the corporate office.

You are on the phone in the job-site trailer when you hear an incredibly loud crash. You run outside and see pandemonium. The cooling tower is on the ground and three workers are down. This accident is happening on your watch and you have got to figure out what to do—quickly!

EMS arrives and you direct the paramedics to the injured workers. Close behind is a reporter who starts questioning everyone with a heartbeat. You call the team leader to find out what to do and are instructed to give a buy-time statement and tell the reporter that more information will be available in 30 minutes. You tell the team leader and spokesperson to get to the site as soon as possible. You need help!

(*Source:* Janine Reid Group, Inc.)

for your scenario. For example, if you developed your scenario for six players and you are expecting 36 participants, set up your training room with six tables with six people per table. If you have an odd number that does not conveniently work out for six people per table, double up two people to work together as one player.

Supply each table with copies of the scenario as well as an envelope containing the player cards. Ask the participants to read the scenario first, then to open the envelope that contains the player cards and each select a card at random. Because the cards are drawn at random, people may play positions for which they have no expertise. For instance, what if a superintendent draws the reporter card or if the president of the company draws the superintendent card? Well, that random element is one of the many advantages of this exercise because it gives everyone an understanding and appreciation for what is expected of the various members of the crisis management team.

Now is the time to put your participants to the test. They should get into the role their player cards describe and work within their individual group to accomplish the following tasks:

1. Determine what action items need to be carried out immediately at the scene of the crisis.
2. Prioritize those action items and determine who should be held responsible for their implementation.
3. Determine what action items need to occur within the corporate office upon notification of the crisis.

4. Prioritize those action items and determine who should be held responsible for their implementation.
5. Prepare the spokesperson for a three-minute interview with the reporter from the table.

The total time allowed to accomplish steps 1 through 4 is approximately 30 minutes. Stay within this limit because you want to simulate the timing of a real crisis as closely as possible, when decisions and direction must be determined very quickly.

At the 30-minute mark, the spokesperson and reporter from each table should simulate a mock three-minute interview in front of the group—one table at a time. Realism can be enhanced by having the reporter accompanied by a technician with a camera and a hot light to simulate the real situation as closely as possible. If you really want to go first class, hire an ex-reporter to participate in your simulation. This addition will give the participants a true taste of the real thing.

Debriefing.   At the conclusion of the interviews, conduct a full debriefing of the total exercise. This debriefing is critical to teaching effective crisis management and reinforces the need for a written protocol. The debriefing includes a review of:

- The action items selected by each team to outline their response strategies at both the site of the crisis and the corporate office
- The assignment of personnel to the action items
- The mock interview

The recommended critiquing method for the first item is to compare Exhibit 3.1, which you must customize to your scenario in advance, to the list that each group develops. Typically, fewer than 50% of the necessary action items will be covered, an outcome that reinforces the need for a crisis management plan. Also, you can underscore the importance of these action items by pointing out that a crisis is dynamic and that successive events will develop differently based on the decisions and actions taken by the team. If the team forgets about an item, the oversight could have a major impact on the outcome of the situation.

Next, critique the interviews between the spokesperson and the reporter at each table. Chapter 5 provides the assistance necessary to conduct this critique, and it should be followed closely because spokespersons tend to say either nothing or too much. They may also fall victim to allowing the reporters to start asking questions, something that should not be permitted during the first hour of a crisis.

At this point, you can play videotapes of people being interviewed during a variety of situations. These examples can show both the brilliance and the stupidity of a wide variety of spokespeople. You can gather these clips simply by recording the nightly news or clipping articles out of newspapers or magazines.

Introducing Your Crisis Management Plan.   The last step in your training module is to roll out your plan and describe its contents. Now, you may wish to introduce the crisis management plan prior to the simulation because the plan can guide the participants toward the correct response strategies. This teaching method is certainly a good one, and its use rests with you, the instructor; however, participants will more readily support the plan *after* they have floundered through a simulation without any such guidance. Floundering emphasizes the true value of the plan.

As discussed in Chapter 3, your crisis management plan should be distributed to only a limited number of people; therefore, the recommended way to walk the participants through each section of the plan is either through a Power Point presentation or overhead transparencies. This teaching format allows you the time to review each section and solicit input from the participants.

Tell all employees where they can find a full copy of the crisis management plan should they need it during a crisis. Also, provide employees with a laminated card (business-card size) that lists the names and phone numbers of the team leader, spokesperson, and their backups. Employees should keep these cards with them at all times.

How long should a training session last?

The time for a training session varies based on the number of attendees. Interactive training can take more time than a lecture does simply because feedback is received and recorded by the instructor. If you use the workshop previously described and invite 15 people, you can complete a training session in two to two-and-a half hours. However, if the group is expanded to 50 people, at least three to three-and-a-half hours should be allowed. A good rule to follow for interactive training is the larger the attendance, the longer the session.

How often should employees receive training on this topic?

All employees should be trained or receive refresher training at least once a year. The core and project crisis management team members should receive an additional training session that focuses on working with the news media because these team members are the folks who will be on the front line and, perhaps, on the front page.

What are some ideas for providing in-house media training?

Working with the news media, especially during a crisis, is a skill—a learned skill. To retain and hone that skill, the spokesperson and backup as well as the core crisis management team should be exposed to frequent training in this area. This will help them practice the art of buying time and delivering effective news conferences. Here are ideas to start you in the right direction.

Buy-time exercise

Everyone on your crisis management team should understand how to deliver a buy-time statement effectively. To test this process, present a very sketchy scenario of a crisis that has just occurred and give a participant only two minutes to prepare for an interview with a reporter who has already arrived on site. At this point, select someone to play a reporter who fires questions left and right and is downright obnoxious.

Here is an example of a sketchy scenario:

> You are the general contractor on an eight-story commercial office building in a major metropolitan city. The structural steel is up to the sixth floor.
>
> Four beams are being lifted to the sixth floor when a crane failure occurs. All four beams fall, bringing down two ironworkers and hitting two people on the ground. One of those people is a pedestrian and is presumed dead. One beam is resting in a major intersection.
>
> The site is in chaos and a reporter is approaching you for a statement. What will you say?

When the two-minute preparation time is up, stage an interview between the participants acting as the spokesperson and the reporter. The spokesperson should deliver a buy-time statement and not entertain questions. If the spokesperson does not accomplish this mission, stop the exercise and review the section in Chapter 5 that discusses the techniques of a buy-time statement. Then repeat the exercise until the participant masters the protocol.

Next, stage a mock news conference. Set up the room with a standing podium and select three to four people to play the parts of the spokespeople for the company in crisis. The remainder of the participants play the intended audience. Here is a sample news conference scenario:

> You arrive at the job site early to find a contingent of about ten Earth First protesters carrying signs and chanting at the construction trailer. The protesters say that they want to stop construction on what they view as an unnecessary project. Signs say the project will wipe out a species of bird and ruin wildlife habitats.
>
> Your superintendent informs you that at least three protesters are staging a sit-in directly in the path of heavy equipment and are refusing to move. He also informs you that the batteries from the trucks, front-end loaders, and graders have been removed and that it will take at least an hour to replace the missing batteries.
>
> Members of the community and a group of reporters are waiting at your site for your response. You elect to hold a news conference and deliver information at one time to both audiences. What will you say?

In this scenario, your spokespeople get 20 minutes to prepare for a mock news conference that includes a five- to ten-minute presentation of information and then a ten-minute question and answer session. Refer to Chapter 7 for direction.

In the meantime, the remainder of the participants prepare to play their parts as members of a community group and as reporters. They should come up with questions for the members of the panel that will challenge their positions. Some of those questions might include the following:

1. If a project provided you with a healthy monetary gain, would you take it on if you knew it would harm the environment?
2. Will this disruption put you behind schedule and, if so, how will you make up for the lost time?
3. Aren't these people just one big pain? Wouldn't it be easier just to have the authorities haul them away?
4. How do you intend to communicate with these people?
5. I am with the Audubon Society. Will you make a donation to our organization to show your support of our cause?

As you can imagine, this scenario can present some interesting dialogue, which is the goal of a simulated news conference. Take the time to develop media-training exercises customized to your company and get some feedback from your colleagues. You will be rewarded for your effort.

Once again, training should occur at least annually because there is absolutely no substitute for trained personnel when a crisis strikes; therefore, a crisis management plan must be supported with an aggressive training program to have value. A well-trained team executing a well-considered plan will help you get off of the front page and back to what you do best.

Chapter Summary

- Frequent training increases the confidence of the crisis management team as well as that of the company's employees. Confident performance is the key to successful crisis management.
- The core crisis management team as well as the project crisis management team should receive training at least once a year. Additional media training should also be delivered once a year, or more frequently, for the spokespeople and backups.
- Incorporate a crisis audit into the training sessions to teach the process of crisis identification. Remember to teach the process of identification so prevention can follow.
- Include an exercise in what-if analysis to enhance the participants' abilities to anticipate the unknown.
- Simulation-based training is recommended because it enables the participants to achieve a level of confidence in their decisions that is not otherwise attainable. Simulations also point out weak areas in both the crisis management plan and the performance of the crisis management team. Corrections can be made before a real event occurs.

- Hire a good mock reporter for your simulation. Not only will your media interviews be realistic but also you will be able to give trainees a flavor of the stories that will result from their interviews.
- Media training should include scenarios that the spokespersons could encounter in real life. Be sure to include a buy-time exercise, statements that provide updated information, and a news conference role-play.

11

CRISIS RECOVERY

When the initial rush of a crisis is over, the crisis management team members will have a strong desire to put the incident behind them and return to their real jobs that make real money. This desire certainly sounds reasonable; however, a few things need to be accomplished before the company can return to business as usual.

The words *crisis recovery* are easy to write, but the actual process of recovery requires reflection and action. Let's talk about the reflection process first.

How do you analyze your company's response to a crisis?

Memories are short—especially unpleasant memories. Therefore, conduct a debriefing session with the members of the crisis management team within one week of the outset of the crisis. To prepare for this debriefing, ask each team member to evaluate the company's response to the crisis. To facilitate this evaluation, a post crisis questionnaire should be completed. Exhibit 11.1 provides a sample format for a postcrisis evaluation questionnaire. The desired result of this evaluation is an honest critique of the performance of the crisis management team during the incident as well as outside resources that were used. In addition, a full review of the crisis management plan should be conducted to determine its effectiveness.

The completed evaluations should be returned to the team leader for his or her analysis of the overall results. At this point, a meeting should be scheduled to debrief the results. All of the members of the crisis management team should participate in this debriefing exercise, as should any outside resources that provided assistance during the crisis. Outside resources might include specialty consultants, agencies, other businesses, and so forth. These individuals/groups can supply valuable input to the evaluation process and may also become loyal supporters in the recovery process.

EXHIBIT 11.1 Sample Postcrisis Evaluation Questionnaire

The goal of effective crisis management is to learn from our experience. To that end, please be honest and thorough with your answers to the following questions so that we can strengthen our existing program.

Your name_____ Date_____

Your role on the team _____

Were you notified in a timely manner? If not, how can the notification system be improved?

_____

_____

On a scale of 1 (poor) to 10 (excellent), how would you rate the way the company managed the incident? _____

What were the company's weaknesses? _____

_____

_____

_____

What are your recommendations for improvement? _____

_____

_____

_____

What were the company's strengths? _____

_____

_____

_____

What, if any, changes should be made to the crisis management team? _____

_____

_____

_____

Should additional training for members of the crisis management team be offered? If so, what type of training should be conducted? _____

_____

_____

_____

EXHIBIT 11.1 Continued

Does the crisis management plan need to be improved or revised? _____

If so, what recommendations would you make?_____

_____

_____

_____

On a scale of 1 to 10, how was our communication with our employees? _____

If your answer is below a 7, what improvements would you recommend?_____

_____

_____

_____

On a scale of 1 to 10, how was our communication with all of our audiences? _____

If your answer is below a 7, what improvements would you recommend?_____

_____

_____

_____

If there was media coverage, was the crisis reported in a balanced fashion? _____

If not, what needs to be done to communicate our concerns to the media and/or our audiences? _____

_____

_____

_____

If there was media coverage, did the spokesperson relay the company's key message(s) effectively? Please provide an evaluation of the performance of the spokesperson._____

_____

_____

_____

Who should be thanked for his or her assistance during the crisis, and what form should that thanks take? _____

_____

_____

_____

continued

EXHIBIT 11.1 Continued

What suggestions would you make to prevent a reoccurrence of this incident? _____

_____

_____

_____

Do you have any other suggestions? _____

_____

_____

_____

_____

_____

(*Source:* Janine Reid Group, Inc.)

Now it is time to move to the action side of the recovery equation, where the team leader's responsibility is to document the group's recommendations and develop a list of action items to effect the agreed-upon changes. In an effort to develop this list of action items, the crisis management team must elaborate on some of the questions asked in the evaluation:

1. Should the existing crisis management team remain intact? If not, what are the recommended changes?
2. Should additional training be required in lieu of changes? If so, what type of training should be incorporated, who should receive the training, and how frequently should the training occur?
3. Was the crisis management plan used by the team? If not, why not?
4. Which sections of the plan were most useful? Which sections of the plan were least useful?

A word of caution: One of the primary targets for discussion is the performance of the crisis management team. Remember that even the best training program cannot prepare an individual for the high level of stress in a real crisis. Furthermore, nothing brings out the best or worst in a person faster than a stressful situation. For example, a member of your crisis management team may deliver a stellar performance during a training exercise, but the stress of a real event may cause him or her to fall short of the team's expectations. The postcrisis evaluation is not a time to demote or terminate that person but rather to seek additional training in his or her assigned area of responsibility or to change his or her position on the team. If the stress proved too much for this individual, the issue may take care of itself because the team member in question will most likely withdraw from the team. In contrast, a team member who appears a bit shaky during a training exercise may deliver a performance beyond everyone's expectations.

The answers to these questions will provide the direction needed for the team leader to enhance the existing crisis management program. Any change should be implemented as quickly as possible after the debriefing, while the topic is still fresh in everyone's mind. As mentioned earlier, a crisis is an excellent teacher, but its students must focus on the lessons learned from their experiences and apply that knowledge to future situations. Otherwise, history will repeat itself.

How do you rebuild your image?

A company's reputation is fragile and can be threatened by a crisis. If the damage is severe, a long time may be needed to restore its reputation and position in the marketplace. However, steps can be taken to begin the restoration process:

- Determine which of your audiences were affected by this crisis and make certain that your communication with them is prompt and that your message is consistent in reinforcing your position. Your audiences must be assured that you are doing everything in your power to make things right. Always include your employees as one of your most important audiences.

- Communicate your concern to the applicable publication or station if you feel as though you were misquoted or misrepresented by a reporter. (Refer to Chapter 5 for suggestions on how to address this situation.) You might also consider writing a letter to the editor stating your side of the story.

- Express your appreciation to individuals, agencies, community groups, businesses, and so forth that were particularly supportive during the crisis. Actively solicit their critique of your company's performance and seek advice on how to improve it. If their confidence in your company is shaken, ask for their recommendations on what you can do to help restore their trust. You may also wish to secure an ad in your local paper to thank those who provided assistance. To circulate your message further, consider obtaining reprints of the ad to mail to all of your audiences.

- Deliver presentations at your local association meetings to share your experience and the lessons you learned from it. This form of education is both cathartic for the company and endearing to your various audiences.

- Be prepared to be asked about the crisis—even long afterward. In addition, all employees must know the answer to give when queried about the incident or the person to whom the question should be referred. The company must speak with one voice until the end of time.

- Begin a public relations program or accelerate the existing one to get your message(s) out to all of your audiences. Refer to Chapter 9 for ideas on this enterprise.

- Implement procedures to make certain this crisis never repeats itself. On a consistent basis, increase your training programs and stress risk awareness with all employees.

Will it ever be over?

Probably not; however, the time a crisis lingers varies, and much of the aftereffect depends on whether you were the victim or the culprit. If you were the victim, the association between the crisis and your company fades within a short time. However, if you are the culprit, a crisis can persist for a variety of reasons:

1. Your competitors have long memories, and they will not hesitate to remind potential clients about your incident.
2. Clients may hesitate in securing your services if they feel as though the same thing could occur on their projects. Prospective clients may also feel uncomfortable working with you if they believe that your company mishandled the situation. Remember, clients' reputations are also at stake during a crisis, and guilt by association is something they want to avoid at any cost.
3. Reporters may elect to exhume your crisis on its anniversary and do a story that breathes life back into all of the ugly details.
4. Employee morale can take a nosedive as a result of some or all of the above. In tight labor markets, such a morale problem could spell trouble.
5. If your crisis received media attention, the media's archives will house the information for a very long time. Should you have another crisis, reporters will not hesitate to bring your past to light.

As an effective crisis manager, your goal is to expect the unexpected and to be prepared for a past crisis to rise like a phoenix again.

How is XYZ recovering?

Return to our hypothetical crisis with XYZ Construction Company. Imagine one week has passed since the cooling tower fell at the medical office building site. The post-crisis evaluation questionnaires (Exhibit 11.1) were completed by the crisis management team. The team leader summarized the findings and called a meeting to review the evaluation. Here is the group's consensus of the company's performance:

- The general feeling was that the notification process worked very well. The team used the emergency contact list (Exhibit 3.2) and made the appropriate phone calls.
- The overall rating on the company's performance was a seven out of ten. The action items on the First-Hour Response Checklist (Exhibit 3.1) were implemented immediately upon notification of the accident and proved to be useful.

However, the project manager mentioned that the project engineer and super-intendent panicked at the outset of the crisis; nevertheless, the checklist and their training allowed them to regain control quickly.

The project manager recommended adding a line item to the checklist that would outline the steps necessary to secure a site after an accident. The group agreed that this new line item was key and requested that the project manager submit his recommendation to the team leader.

The team also felt that the mechanical subcontractor took too long (four hours) to notify the families of its two injured employees. However, the team leader argued that, as the prime contractor on this project, XYZ's responsibility had been to make certain that notification of the next of kin was made as quickly as possible—no matter who employed them. (Refer to Exhibit 2.2 for clarification of "Who is in charge.") True, four hours exceeds acceptable boundaries, but XYZ should have notified the families before then.

Fortunately, the subcontractor's two injured workers have returned to the job site and have been assigned light work. XYZ employee Larry Lynn, who had a skull fracture, is in good condition and is expected to be released from the hospital tomorrow. XYZ has done a tremendous job of standing by the families of the injured and will continue to do so.

- The crisis management team, in general, performed admirably. Unanimously, the team decided that a companywide debriefing and refresher training session should be scheduled within the next two months, while the incident is still fresh in everyone's mind. This session will provide a perfect forum for employees to discuss the strengths and weaknesses of the company's response to the accident and contribute their recommendations on how XYZ could perform better should another incident occur.

- XYZ felt its communications with its employees ranked a ten. XYZ is fully aware that its employees are the company's voice to the outside world. Therefore, the crisis management team made certain that the lines of communication remained wide open throughout the crisis.

- Communications with XYZ's audiences were voted less than complimentary by the crisis management team. The group ranked this area as a five. Generally, team members felt that everyone became so caught up in the rush of the crisis that no one was assigned the responsibility of communicating XYZ's side of the story to its key audiences. As a result, XYZ was pulled off one short list of possible contractors for a construction project and the industry rumor mill was running at full tilt to produce unfavorable stories about XYZ. In addition, the media was having a feeding frenzy during the first 24 hours of the crisis while information was very scarce and speculation reigned supreme. However, such unfortunate events did not excuse XYZ from contacting its audiences to provide status updates and assurance that XYZ was in control of the situation.

Unanimously, the team felt that the responsibility of identifying and communicating with key audiences lay with upper management. One team member suggested that a category entitled "upper management" be added to the First-Hour

Response Checklist and that notifying audiences be among the responsibilities listed in this category. Furthermore, upper management should be assigned the responsibility for making certain that the next-of-kin notification is expedited. The group asked the team leader to modify the checklist to accommodate these changes.

Also, the team noted that special attention must be paid to community relations—on all projects. XYZ thought it was doing the right things, but, as became obvious, it was not sensitive enough to neighbors' concerns. XYZ should have done more research in this area. The team requested that the marketing department develop a plan of action for community relations. The goal is to make certain that neighborhoods are friendly at every XYZ project.

- The media coverage, on balance, was fair. This coverage was due largely to John Smith's excellent job at communicating the company's key messages consistently. He made certain that verifiable information was released in a timely fashion and that requests for interviews were addressed promptly. The group did note that the media sensationalized the event during the first 24 hours when facts and details were unavailable; however, John made certain that inaccuracies were corrected in both his verbal interviews as well as written followups.

- The crisis management team agreed that additional media training should be required of field personnel as soon as possible. The superintendent, who gave the buy-time statement, did not stick to the script and started answering a reporter's questions. (As we discussed in Chapter 5, the last thing you want to do in a buy-time statement is to entertain questions from a reporter.) Luckily, the superintendent excused himself before a troubling question was presented.

- Collectively, the team identified the people and organizations that provided assistance to XYZ during the crisis. The team agreed that the president of XYZ should make contact with each of those involved and thank them personally for their contribution. This should occur within the next few days.

- The issue of Mary Smith's being harassed by the workers on the site was raised. The team was not pleased to discover that the director of human resources had just now gotten around to scheduling a meeting with Mary Smith. XYZ should have responded immediately to her complaint. However, interviews were conducted with the employees on site and the director of human relations will copy the crisis management team on the findings of all interviews.

On balance, all the team members felt that they did a fair job in handling the crisis but that they had a lot of room for improvement. A follow-up meeting was scheduled for the following Thursday, when Rick Kearney, the director of safety, felt that the preliminary findings of the investigation would be known.

The investigation

Rick Kearney was hired by XYZ just one month ago. He is the only full-time safety person on the XYZ payroll and is feeling overwhelmed right about now. Rick left the meeting with the crisis management team and went directly to the job site to complete some paperwork. Upon arrival, he met the OSHA investigator, who was just leaving the job-site trailer. A polite exchange took place, and Rick entered the trailer and went directly to his desk. While shuffling through papers, Rick noticed a memo from the OSHA investigator to the OSHA assistant area director. He presumed that the investigator inadvertently left the memo behind. Rick read the memo and returned to XYZ's office to meet with the team leader. Shaken, he tried to recall the highlights of the memo.

- One of the nylon chokers being used to secure the load to the hoist line of the crane was found to be badly worn. The red safety threads were exposed in several places.
- Another choker was the wrong size for this load. The rated capacity of the choker was about 20% of the weight it had to hold.
- Employee interviews revealed that the load was being lifted over a staging area. One of the injured was working in this staging area at the time of the accident.
- Interviews with three workers revealed that they did not speak English and that they had never received any training that they could understand.

Rick's head was swimming. The investigator who wrote the memo indicated that he was leaning toward assessing both XYZ and the mechanical subcontractor with the maximum penalty. The memo ended with a brief sentence stating that the investigation should be completed within the next ten days.

The team leader contacted the president and vice president of XYZ and shared the disturbing news. This information dealt a blow to the management team because XYZ had not wavered in its commitment to safety; however, these findings would raise doubt among XYZ's audiences. The team had to work fast to determine a course of action.

The group decided that XYZ must be prepared for all possible citations that OSHA might issue. Feeling bad about what happened was not good enough—XYZ had to demonstrate a course of action to ensure that it would do everything in its power to prevent another occurrence. To that end, a full-time safety professional would be hired to assist Rick Kearney in making XYZ's safety program significantly more thorough.

Rick was perplexed about the language issue, as he made certain that the safety meetings were delivered in both English and Spanish. But he had been with XYZ for only a month, so the "miscommunications" could have occurred before his time. Rick planned to talk to the subcontractors on the project where the crisis occurred to make certain that everyone received training in a language he or she could understand.

The group determined that a full-scale communications effort must be put into place with all of XYZ's audiences. In Chapter 6, we discussed the fact that you are judged more on your response to a crisis than the actual crisis itself. Therefore, if XYZ was found at fault when the OSHA findings were released, company officers would be well served if they stepped up to the plate, admitted the problem, and communicated how they would correct it. Then they would have to follow through with their commitment. To do otherwise would lengthen the recovery period.

As we discussed in previous chapters, this crisis will never totally go away. Therefore, the company must always be prepared to position itself should it be questioned by any of its audiences. For instance, let's fast-forward six months to when XYZ is one of three contractors being considered for a big project with a highly visible owner. During XYZ's final presentation, the owner states that he is nervous about hiring a general contractor with a bad safety record. At this point, XYZ must make certain to communicate its key points quickly and succinctly. Those key points must include an acknowledgment of the tragedy of the incident and the steps XYZ has taken to make certain that it never happens again.

Chapter Summary

- Resist the temptation to put the crisis behind you before a full evaluation of the company's performance can be completed by the crisis management team.
- Consider inviting outside resources that were engaged during your crisis to the debriefing. An outsider's reflection on your crisis can provide valuable input to the recovery process.
- A postcrisis evaluation questionnaire should be completed by the members of the crisis management team within one week of the outset of a crisis. The resulting recommendations for improvement should be implemented immediately to ensure a stronger crisis management program.
- Do whatever it takes to make certain that the crisis never happens again. Begin by increasing your company's crisis management training.
- Quick action and a consistent message are the keys to rebuilding a company's image after a crisis.

12

WHEN BAD THINGS HAPPEN
TO GOOD COMPANIES:
TWO CASE STUDIES

The best way to begin to bring this book to a close is by sharing two stories of actual crises. In analyzing these case studies, we look at the actions taken by the companies and compare them with the material covered in previous chapters.

I would like to extend my gratitude to these two firms for allowing me to interview their employees and dig into their closets. I have a great deal of admiration for both firms.

Walsh Construction Company

Company Background

Walsh Construction Company, founded in 1961 and located in the Northwest, is a midsize general contractor specializing in commercial and residential multifamily construction, including both new construction and renovation work. Walsh is headquartered in Portland, Oregon, and has offices in Seattle, Washington, and Davis, California. Walsh has a reputation for high quality with its public and private owners and enjoys a strong base of repeat clients.

Walsh installed a crisis management plan one-and-a-half years prior to the event about to unfold. In addition, Walsh provides its employees with frequent training on their roles and responsibilities in the unlikely event of a crisis.

A Bit of History

Founded in 1851, Portland, Oregon, has grown into a major metropolitan city of approximately 1.7 million people. Like other cities across the country, Portland has contended with suburban flight, the movement of businesses away from the central city to industrial parks located in the suburbs. This flight has left behind large numbers of empty multistory warehouses and commercial buildings. In the

1980s, taking a cue from their counterparts in cities such as New York and Chicago, developers saw the possibilities of converting the empty buildings into residential and multiple-use spaces, including loft apartments and shopping centers.

A relatively recent discovery for developers was the neighborhood now known as the Pearl District, defined as the area of northwest Portland connecting downtown to upper northwest Portland. Until ten years ago, the area was predominantly industrial; today the area is home to a diverse mix of people and an eclectic array of restaurants, bars, coffee shops, art galleries, and other businesses. The area became known as the Pearl District as the old buildings were transformed into habitable and usable spaces, often revealing significant architectural surprises within.

Walsh Construction Company began working in the Pearl District when it converted the seven-story McKesson Robbins Warehouse into 78 loft apartments. Since that time, Walsh has been involved in the transformation of more than seven buildings, with other projects in the works.

The Project—The Crisis

On October 7, 1998, Walsh began work on a five-story apartment complex in the Pearl District named The Kearney Plaza Apartments. The project had a wood-framed construction and would house 131 apartments, a retail complex on the first floor, and a basement garage of 48,000 square feet.

On August 18, 1999, The Kearney Plaza Apartments were 80% completed. At 4:00 P.M. the crews were done for the day. Everyone packed up the gear and headed for home. Ken Bello, the superintendent, secured the site and drove to the office to complete some paperwork. From there, he called it a day.

At 4:10 A.M. on August 19, 1999, the Portland Fire Department received a call from a resident of a condominium building in the Pearl District. The caller said that The Kearney Plaza Apartments were on fire and that the fire appeared to be out of control. At 5:00 A.M., Ken Bello was awakened by a knock on the door from his neighbor, who had heard about the fire on the morning news. In shock, Ken went to the job site and phoned Bob Fisher, who was another Walsh superintendent on a project located just eight blocks from The Kearney Plaza Apartments. Ken yelled into the phone, "My building is burning down!"

The message did not register with Bob because he was still half asleep, so he yelled a resounding, "What?"

Ken shouted back, "I said my building is burning down!"

Bob screamed, "Just what are you trying to tell me?"

This exchange continued until Bob finally woke up and realized that *his* job was not burning, but *Ken's* was.

> Let's take a timeout here. In retrospect, this telephone call can be viewed as a humorous situation; however, as we discussed in Chapter 2, people have a wide range of responses to stress. Now, Bob was just waking up—but the information was slow to get through. This response is common even for people who are wide awake!

Bob Fisher immediately called the team leader, Bob Forster, Walsh's executive vice president, who put the crisis management plan into effect and began to mobilize the team. Phone contacts were made through the emergency contact list (Exhibit 3.2) and the action items accomplished as indicated in Exhibit 3.1.

An interesting side note here: Bob Walsh, the company president, was out of town at the time of the fire. Walsh, who has a high profile in Portland, was the designated spokesperson in the event of a highly visible crisis. He was notified of the fire immediately by telephone, but he expressed confidence in his team and did not feel his presence was necessary. Nevertheless, he did remain in constant contact with the team by phone.

Mobilization

Walsh's crisis management team rose to the occasion. Bob Forster was at the scene by 5:30 A.M., closely followed by Michelle Potter, Walsh's safety director. The fire was now classified as a five-alarm fire and 125 of the city's 164 firefighters were on the scene. The firefighters concentrated on keeping the fire from spreading to the surrounding occupied buildings.

The obvious visual impact of this fire created a lot of media attention and news coverage began at 5:00 A.M. Randy Boehm, principal for Walsh, assumed the responsibilities of spokesperson. In the beginning, reporters were interviewing people standing on the street. At 11:00 A.M., Randy had his first live interview on site, which turned out to be his most challenging media interaction.

The Interview

The reporter introduced himself and then asked Randy if Walsh had ever had a similar occurrence. Randy was prepared for this question and explained that Walsh had a fire on another project in 1982. The reporter mentioned that if Walsh did not come totally clean, he would dig up all information he could on the company. Randy assured him that this was the only incident.

Remember, reporters can access databases that will give them information on your past. This access includes citations issued by governmental agencies as well as any past media attention. In Chapter 3 we talked about the importance of knowing what is in your closet and understanding how to position those situations should the need arise.

The reporter did his homework and determined that Randy was telling the truth and not hiding anything. Randy gained a lot of credibility with that particular reporter; however, the story does not end there.

Michelle Potter and Bob Forster were still on site, and the media figured out who they were and started asking questions. Michelle did a tremendous job not only of handling the safety aspect of the job but also of acting as Bob Forster's team coordinator throughout the crisis. Michelle also protected Bob from reporters by politely referring them to Randy, who was at the office.

> The decision was made to keep Randy at the office for two reasons: (1) he could handle the media calls from one location and (2) he could coordinate all of the internal actions that needed to take place from a corporate stand-point (see Exhibit 3.1). Randy wanted to make certain that the employees were kept up to date at all times and that everyone remained calm.

Randy received about a dozen media calls the first two days of the crisis. Obviously, the cause of the fire was not known at this time; however, questions regarding the wood-frame construction were at the top of all of the reporters' lists. Apparently, the firefighters were vocal with the reporters about the dangers of wood structures being built to the height of five stories in a downtown area.

> As a bit of background, the danger of fire in wood-frame buildings is substantially reduced once the drywall is installed. Also, the fire protection equipment that was required of The Kearney Plaza Apartments was extensive. Fire sprinklers were installed in each of the apartment's rooms. The fire protection equipment was installed before the fire broke out; however, the city had just stubbed the water to the building the day before the fire and the lines were not yet charged.

Randy did his homework and anticipated questions relative to wood-frame construction. He developed organized and succinct responses that reviewed the cost-effectiveness of wood-frame construction, which in turn offered affordable downtown housing. Randy also mentioned that the fire danger was minimized once the drywall was installed. He added that Walsh Construction Company wanted to work more closely with the City of Portland to get the water stubbed into buildings earlier in the construction schedule so that the fire protection equipment could be activated sooner.

Walsh thought more such land-mine questions could be lying in wait, so it contacted Bonnie Gilchrist of Gilchrist & Associates, a Portland communications company, to brainstorm other questions or concerns that might arise. Walsh felt that a question about the insurance coverage might surface soon, and that presumption proved right.

During one of Randy's interviews, a reporter suggested that if the insurance would cover everything, the fire was really no big deal to either the owner or to Walsh. Further, the reporter speculated that the workers' being without tools and a job did not matter either. As you can imagine, that speculation was like a slap on the face, and both Ken and Randy jumped on it. They insisted that the fire was a tragic event to everyone involved, that the workers' tools would be replaced, and that Walsh would find them work on another project.

> As we discussed in Chapter 5, reporters are always looking for an angle and a culprit. Thus, the spokesperson must control his or her responses and set the

record straight when necessary. The key, obviously, is in anticipation and preparation.

As Randy reviewed the television and print coverage of the crisis, he felt that the overall reporting was fair and balanced. However, Randy was prepared for even the ugliest of questions and did an admirable job in presenting Walsh Construction Company in a positive light. Speaking from experience, I know that accomplishment is not easy.

Communication

Bonnie Gilchrist also advised Walsh to meet with its employees frequently to keep them apprised of the situation. Walsh has a reputation for being a loyal employer and its doors of communication are always wide open. Hence, Walsh's employees showed their support throughout the crisis.

Then came the issue of contacting Walsh's audiences (Chapter 8). This step was one of the few places where Walsh felt it fell short; its marketing database was less than adequate for getting a letter out in a prompt and organized fashion. This was quickly rectified, however, and a letter of information and reassurance went out to all major clients by 5:00 P.M. on the day of the fire.

Four Days Later

The fire department took four days to make sure that all of the hot spots were out and that everything was under control. The media attention diminished and the focus turned to the investigation.

As Walsh completed its postcrisis review, company officers were pleased to discover that the owner of the project and their insurance broker were extremely impressed with Walsh's prompt and comprehensive response to the crisis. As a matter of fact, Walsh began construction on the new Kearney Plaza Apartments on October 7, 1999. Ironically, October 7 was exactly one year from the start of the initial construction.

On September 30, 1999, the Portland Fire Bureau announced that the cause of the fire was arson and that a reward was being offered to anyone who could lead authorities to the culprit.

What Would Walsh Do Differently?

Randy listed three areas that the company needed to improve:

1. Crisis management team members would modify their First-Hour Response Checklist (Exhibit 3.1) to include additional action items. One of these would be to have two team leaders, one at the office and one at the job site. Randy felt that an organizational person at each location was needed to facilitate the coordination of activities and communications among the company's various audiences.

2. Michelle Potter was carrying a tremendous amount of responsibility in both implementing the company's safety program and acting as Bob Forster's team coordinator. Michelle needed more support from people who normally did not take directives from her.

> In Chapter 3, we talked about the importance of not sticking to the company's organizational chart because it would vanish in a crisis. Well, that advice is easy to say but hard to do, because corporate politics can come into play.

In response, the crisis management team decided to hold a companywide meeting to discuss the roles and responsibilities of the team with all employees and to solicit their support for the team members should another crisis occur.

3. Walsh's marketing database failed during a time when it was sorely needed. Immediate plans to update the database were already in the works.

Randy claimed that the crisis management plan did its job. More importantly, the crisis management team performed admirably. The training paid off. It even compensated for the absence of key Walsh personnel. Not only was the company's president out of town, but one of its vice presidents was also away at another site and its general superintendent was in the hospital. Despite these absences, all team members gave 110%, and the owner's vote of confidence confirmed the effectiveness of their efforts.

Asphalt Products, Inc.*

Company Background

Asphalt Products, Inc., (API) was headquartered in Benica, California, and had three plants within a 100-mile radius of Benica. Founded in 1976, API was family owned and produced a variety of products using asphalt cement as the raw material.

Product Background

Asphalt cement is a noncombustible, nonhazardous thermoplastic material used primarily for the construction of roads. It is also used for roofing material, the lining of ponds, sewage lagoons, potable water tanks, fish hatcheries, and other waterproofing and adhesive applications. It is a liquid material when heated above 250 degrees and a solid material below this temperature. It is the glue that holds roads together, seals out moisture, and fills the cracks. Today, billions of tons of asphalt cement hold U.S. roads, highways, driveways, parking lots, roofs, and runways together with aggregates. Ninety percent of the roads in the United States are constructed using asphalt cement.

Author's note: This case study is true; however, the name of the company and the locations of its plants have been changed in view of ongoing media attention.

The Leyden Plant

The Leyden plant was in operation since 1993 and, as with all of API's plants, was in full compliance with all of the regulatory agencies. The area surrounding the Leyden plant was zoned for commercial use in 1984. Periodically, the plant received complaints from area residents about the odors it emitted. These odors came from off-gas emissions caused by the blending of raw asphalt and polymer additives.

In response to the neighborhood's concerns, API joined forces with a federal energy and environmental laboratory to develop a biofiltration technology. This technology involved an odor-suppression system consisting of three separate granular-activated carbon (GAC) beds to treat the plant's primary emission sources. API installed a biofiltration system at its Leyden plant in the summer of 1996. The results were impressive and the biofilter removed much of the odor. As a matter of fact, this technology was the focus of an article published in a trade journal.

One year after the installation of the biofiltration system, API circulated a survey to the residents of the Leyden area to confirm the company's findings. Exhibit 12.1 presents the results of the survey. Almost half of the respondents reported that the odor conditions had improved since the installation of the biofiltration system.

EXHIBIT 12.1 1997 Odor Survey Results

Questions/responses	% of respondents
What description below describes where you live?	
Northeast of the plant	26.2
Around the plant	39.6
Southwest of the plant	25.1
Other	9.1
How long have you lived at the residence indicated above?	
1 year	3.2
2 years	1.6
3 years	1.1
4 years	1.6
More than 4 years	92.6
Have you ever smelled odors that you believed originated from the asphalt processing plant?	
Yes	77.4
No	22.6
Were the odor conditions better in 1997 than in 1996?	
Better	48.8
No change	36.0
Worse	15.2

(*Source:* API)

A Neighbor

Sue Jones and some other local residents had opposed the Leyden plant for quite some time for a variety of reasons. One resident claimed that she had developed respiratory problems as a result of the plant's emissions. She also claimed that the value of her home had declined because of the smell, the excessive truck traffic, and the possibility of fire from the flammable materials stored at the plant.

Sue Jones had become the neighborhood group leader and was spearheading an effort to file a petition with the county to prevent further expansion of industrial zoning in the area of the Leyden plant. As part of her research, Ms. Jones visited the assistant plant manager of the Leyden plant in July 1999. Her demeanor was cordial but concerned. She was concerned about the odor, the truck traffic, and the types of material stored on site. The assistant plant manager spent a great deal of time educating Ms. Jones about the technology that API was using to minimize the odor. He also reviewed the routing system for the trucks to minimize noise. Finally, he explained the content of the materials used to produce asphalt cement. All of this information seemed to relieve Ms. Jones of her fears, and she even admitted that she was unaware of the steps that API was taking to be a good neighbor.

> Let's take a short timeout and review Chapter 1, where we discussed the importance of identifying potential crises before they become real ones. Now, a Monday-morning quarterback might say the assistant plant manager should have seen this crisis coming; however, analysis of this case study has many lessons to teach. Should Ms. Jones's visit have been recognized as a warning of future problems? Such a call can be difficult to make, but let's read further to see what happened.

Shortly after her conversation with the assistant plant manager, Ms. Jones contacted a television reporter and arranged to meet her at the front gate of the Leyden plant. She then proceeded to air her grievances and concerns about zoning and plant emissions. This interview surprised API and, needless to say, the media coverage was one-sided and less than flattering.

> Another timeout. The media's responsibility is to make certain that their reporting is fair and balanced, but this particular reporter made no attempt to contact API to discover its side of the story. This omission is unacceptable, and API had the right to contact that reporter to query her on why the company was not allowed equal time.

Shortly after the television report aired, Sue Jones contacted the residents of Leyden to solicit their participation in a public meeting to discuss zoning issues in the surrounding area. The rumor was that API might be a target for discussion, as it sat on 66 acres zoned for light manufacturing and was recently the focus of a television report.

API got wind of the meeting and debated whether it should attend. The situation was perplexing for API because it had always made a strong effort to be a good neighbor. API was in full compliance with all of the regulatory agencies and it was spending a lot of money on research and development to uphold that effort. What more should the company do? Should it be proactive and attend the meeting to support its efforts, or should it sit back with the understanding that it had done nothing wrong?

After much deliberation, API decided to send a representative to the meeting. Prior to the meeting, however, API would develop a fact sheet to address the positive efforts and contributions that the company had made to the community. Exhibit 12.2 is a copy of this fact sheet.

> In Chapter 9, we discussed the concept of doing good things and getting credit for them. API was facing one of those times when it needed to remind the community of such contributions.

If appropriate, the API representative would distribute the fact sheet at the meeting and answer any questions that might arise. Unfortunately, this intention did not become reality.

EXHIBIT 12.2 Fact Sheet for Asphalt Products, Inc.

Facts about Asphalt Products, Inc. (API)

- API has developed the leading edge technology in asphalt products supplied to contractors, cities, countries, and state road departments for use in maintenance and construction of our transportation system. This technology results in longer lasting and better performing roads.
- API utilizes recycled materials in its products creating a low cost, environmentally friendly alternative for the use of waste plastic and tires rather than disposal.
- Residents of the Leyden community drive on materials produced by API every day.
- Asphalt cement is non-combustible, non-flammable, and non-hazardous. The flash point is between 500 to 600 degrees Fahrenheit. The asphalt cement at the Leyden plant is stored at 335 degrees Fahrenheit.
- API spends almost $1 million per year on supplied and services provided by Leyden area vendors.
- Community involvement and being a good neighbor is a core value of API. We take community concerns to heart through support of local 4-H programs, youth sports, wellness campaigns, state fairs and other fund raisers.
- API is a global leader in asphalt product and environmental technology. API will continue its partnership with a federal energy and environmental laboratory to optimize the biofiltration technology and minimize plant odors. To date, API has invested over $500,000 in this technology at the Leyden plant. In 1997, API conducted a survey among neighbors to analyze the progress of this technology and the results were favorable.
- Asphalt cement is also used for roofing material, pond and sewage lagoon membranes, potable water tank sealant, fish hatchery tank lining and other waterproofing and adhesive applications.
- API is in full compliance with all regulatory agencies
- API employs over 100 people in the Leyden Valley whose livelihoods depend on the success of the Leyden plant.

(*Source:* API)

The Meeting

Paul Jones, regional sales representative and spokesperson for API, attended the meeting with fact sheets in hand. His intent was to listen to the residents' concerns and speak up when the timing was appropriate. Eighty people attended the meeting. The three county commissioners who were invited did not show up; however, the TV reporter who earlier interviewed Sue Jones was in attendance. The meeting began in an organized fashion with an overview of the purpose and objective of the meeting. Initially, the meeting's purpose was to inform those in attendance of the zoning issues, but the objective of the group was to "alter the zoning for the surrounding areas."

Sue Jones indicated that API was exceeding the description set forth in the zoning ordinance and that if API were allowed to continue operations, she feared what would happen next, picturing more facilities and businesses moving into the area.

The discussion centered around the group's frustration with the county commissioners, and API was pulled in as ammunition against them. The group became more vocal as the topics for discussion were expanded. First came talk about how difficult it was to get a building permit to do even the simplest remodeling job. Then, the discussion moved to the effect that the odors from the Leyden plant were having on housing values. Another issue was raised by a woman who complained that her mother could not live in her house any longer because of respiratory problems that she developed as a result of the odors. The claim was that the mother had to leave her own home and move in with her daughter to cure her problem, but now the family was unable to sell the mother's house for a fair market price. Furthermore, the husband of Sue Jones said that he had spoken with the fire department, which told him that if a fire broke out at the Leyden plant, the department would be unable to suppress the flames.

As the temperature rose in the room, the question and answer session began. Accusations were aired about possible payoffs, bribery, and land swaps to allow API permits to construct additional buildings. Other claims were made that API was not fair to the community and that the company did not pay enough taxes.

Paul sensed that it was not in the company's best interest to take this group on alone, so he took copious notes and returned to the office for a debriefing with API's president. While in a brainstorming session, Paul received a message that the TV reporter at the meeting wanted to interview him about API's side of the story. She also wanted to interview Sue Jones for the same segment. Paul agreed to the meeting and immediately began anticipating questions that might be asked.

> In Chapter 5, we talked about the process of getting ready for an interview. This process entails anticipating potential questions that a reporter could ask and then developing a response to each that positions the company in the most favorable light. Paul knew this interview was critical, so he made certain that he was ready for it.

The interview lasted for approximately 30 minutes, and Paul felt confident about his responses. The segment ran for about 90 seconds, during which Sue Jones had more than half of the time to air her views. Nonetheless, Paul felt that the report was balanced.

The Inspection

The following day, API received a visit from the Air Quality Division of the Department of Environmental Quality (DEQ). DEQ had received a complaint about the Leyden plant from an anonymous source. The representative from the DEQ toured the plant for approximately two hours and acknowledged that API continued to be in full compliance with all regulations.

What Is the Next Step?

What should API do to neutralize the situation, and can or should they try? Here lies the dilemma. API has done nothing wrong, but wants to be a good neighbor. Its wish is to set the record straight and educate the community about the company's operations and the production process of asphalt cement. Here are ideas API discussed:

1. Initiate a meeting with Sue Jones and her family at the Leyden plant. The purpose of the meeting would be to educate the family on the production methods of asphalt cement.
2. Secure an ad in the local paper. The ad would draw from the fact sheet (Exhibit 12.2) Paul prepared for the neighborhood meeting and would explain API's purpose.
3. Host a community appreciation day at the Leyden plant and provide food and tours of the plant.

At this writing, the jury is still out about the first step API will take to tell its side of the story; however, the company is leaning toward the idea of taking out an ad. API's progress from that point forward will be dictated by the feedback it receives from Leyden residents.

What Would API Do Differently?

I asked Paul what API would do differently if it had the chance to relive this episode. He replied that the company would like to have resolved the issue before it ever got to the public meeting stage—if that resolution were possible. Perhaps API could have circulated its fact sheet after the first meeting with Ms. Jones at the Leyden plant in July to educate the residents about the plant and reduce their fear of the unknown.

Paul also believes that if API had contacted Sue Jones prior to the public meeting to request equal time, the resident's fear factor could have been minimized or eliminated.

Could this case study be classified as a crisis? No one was hurt or killed. Nothing blew up and nothing shut down. However, the definition of a crisis, as stated in Chapter 1, is "any incident that can focus negative attention on a company and have an adverse effect on its overall financial condition, its relationships with its audiences, or its reputation within the marketplace." Based on that definition, API clearly had a crisis on its hands.

This is an interesting case study because the asphalt supply company was a victim—an easy target for a neighborhood group because of the materials it produces. I have worked with the company and personally know it is concerned about the communities where it works. API is committed to and recognized for developing technologies to eliminate the odors from its plants and for remaining in full compliance with all regulatory agencies.

The company is now in the process of communicating these messages to surrounding communities and to its audiences. API's mission is to make certain that its voice is heard consistently—not only in the area in question but in all of the locations where it chooses to operate. Perhaps the ad is the first step in community and perhaps even media relations.

13

CONCLUSION

Lessons Learned in Crisis Management

I have become a loyal student in the school for crisis management planning in the construction industry. This school is different from any other educational institution because the exams are unforgiving. The all-to-frequent tests are delivered in real time and involve real dollars. One wrong decision can mean the difference between a positive outcome or the uncontrollable opposite.

I have had the opportunity to work on many construction-related crises over the years and would like to share some of the lessons that I have learned.

Create a risk awareness attitude within your company

The ultimate goal of an effective crisis management program is never having to use it. To ensure this outcome, an environment of risk awareness must be part of a company's culture. No one is immune to risk; however, identification leads to prevention, and an alarm sounded early can prevent a possible nick in the company's armor.

Develop a crisis management plan prior to a crisis

Developing a crisis management plan prior to a crisis may sound like common sense; however, many companies feel as though they are immune to negative situations, so when the unthinkable happens, they fall victim to corporate paralysis. In contrast, a company with a crisis management plan can avoid such paralysis because the plan will have prepared the company under siege. This is not to say that a plan will address every possible catastrophe that could befall a company. No plan can accomplish that task, but a well-thought-out crisis management plan provides your company the direction needed to make better and faster decisions during a crisis.

On another note, if your company has done something wrong, a plan will not make it right; however, it *can* help a company make the best of a bad situation.

Keep your crisis management plan current

You may spend weeks organizing your plan, researching information, and obtaining background materials. Do not become complacent and allow your plan to become outdated, because nothing is more frustrating than needing information in the crunch of a crisis only to find out that what is available is no longer valid. Instead, once developed, your crisis management plan should be reviewed and updated quarterly.

Practice your plan

All salaried employees should receive training a minimum of once a year on what to do should catastrophe strike. All must understand that they have the ability to set the stage for a positive outcome or, conversely, to create so much negativity that the company will have an extremely difficult time just breaking even. Training is critical for those times when an individual either cannot find the plan or elects not to use it. At such times, training is the only thing that stands between a company and chaotic disaster. The sports analogy that says "You will play the way you have practiced" describes the way a well-trained crisis management team rises to the occasion. Training also reinforces the critical thinking skills of those on the team and helps them remain focused during a crisis rather than default to emotional responses.

The company spokespersons should receive refresher training at least twice a year to keep skills sharp. They must be fully prepared to stand in front of a group of reporters, with the hot lights in their eyes, microphones in their faces, and questions coming at them from all directions. The feeling that this experience creates can be paralyzing; however, frequent media training reduces much of the anxiety that a reporter can create. Your worst day is a reporter's best day, and an untrained spokesperson is excellent prey for a reporter.

Your actions during a crisis will be closely scrutinized by your various audiences

In a crisis, always maintain open and honest communication with your various audiences because they will all keep a very close eye on your actions or lack of them.

In the event of an employee injury or fatality, notification of the employee's family must occur as soon as possible

I worked with a company that experienced a job-site accident that resulted in a serious injury. As the Flight for Life Helicopter left the site, I approached the owner of the company and said that we needed to contact the employee's family immediately. The owner was not comfortable in making the call until he had a chance to speak with the attending physician. He felt as though he needed some solid information to communicate to the family before initiating the call. Unfortunately, my pleas for reconsideration went unheard and the employee died two hours after admission to the hospital.

Again, practice common sense, put yourself in the position of the family, and determine when you would want to receive notification. If the life of a member of my family were in peril, I would want to know as quickly as possible. To be notified any later is unacceptable.

Witnesses to an accident should be offered critical-incident stress counseling

Research has shown that individuals who witness an accident can experience substantial physiological and psychological effects that can linger for months, or longer, if not addressed. Seek critical-incident stress counseling for the witnesses and the project team within 24 hours of the incident. The goal is to return individuals to work, and a normal life, as quickly as possible.

Do not let a crisis overwhelm you

Easy say, hard do! A crisis can be an overwhelming event to manage, and it will not hesitate to take control of your company—if you allow it to do so. When a crisis strikes, the team leader must take control and logically break the event into manageable pieces. These pieces are then further reduced to a series of action items to be handled by the crisis management team and outside resources. This approach reduces the anxiety caused by an unorganized approach to managing a crisis.

Expect mistakes

No one is perfect in a crisis. The only crime in making a mistake is failing to take action quickly and responsibly to correct the situation.

Closing Thoughts

I have had the honor of working with some of the best companies in the design and construction industry. These companies have supported the thesis for this book; therefore, I felt it fitting to end with their words of wisdom. The testimonials that follow come from professionals in the field who know what a crisis means. Please listen to them.

> We recently developed new crisis-management plans for each of our companies in Skanska USA. The exercise was definitely worthwhile. We now have clear procedures and lines of communication to deal with a crisis. Hundreds of employees were trained and have a working knowledge of what to do and what to say (and not say) in a crisis. That knowledge is the most important benefit of the work we did. Our people were forced to

think about all of the risks that we undertake everyday on hundreds of job sites and offices around the country. I believe this training helps us focus more on the risks and manage our work more effectively, results that in the long run mitigate risks substantially.

—Stuart E. Graham
President/CEO
Skanska USA

Having a crisis-management plan in place on each construction project, regardless of the size of the project, is essential in the construction business. Construction projects are potentially high-profile targets for the news media and for negative public opinion when the unexpected happens. The media, especially, can be fierce if there is any possible basis for sensationalism. Each project should have a tailor-made crisis-management plan that includes the protocol for accommodating any conceivable crisis. A plan must also include a protocol for subcontractors, who typically are not properly prepared for a crisis on the job. Remember, a subcontractor's crisis *is* the general contractor's crisis.

Due to narrow profitability margins currently existing in the construction industry, a mishandled crisis can significantly impact a company's bottom line. One poor job, or even the perception that a job was handled poorly, can affect the positive outcomes and influence of twenty successful ones. Thus, crisis-management plans are vital to curtail or soften the blow of a crisis on the construction project.

—Ronald G. Norby
Vice President
Hensel Phelps Construction Co.

I have had the opportunity to participate on some of the largest projects in the world. All of these projects have had excellent safety programs implemented by highly qualified people; however, "stuff" can still happen. When a crisis occurs, panic typically takes control and people do not think clearly. In this type of situation, a company with a crisis-management plan has an advantage because this protocol will help to ensure that all of the bases are being covered and that the right steps are being taken in a proactive manner. Simply put, it is just good business to be prepared.

—Bradley D. Giles, PE CSP
Group Director of Environmental, Safety and Health Services
Morrison Knudsen Corp.

The first thing I think about when people talk about crises is that 90% of the construction industry is not prepared for them. A crisis does not give you any warning—it just happens. I speak from experience because we were not prepared for our first big crisis, and it was a hard lesson to learn. Since that time, we have made a commitment to be prepared and now feel confident that we can handle a negative situation in a professional manner. An example of that confidence was exhibited during the Great Chicago Flood. When City Hall called for our assistance, we activated our crisis-management plan and were able to act proactively.

—John Kenny, Jr.
Vice President
Kenny Construction

Crisis management is something no one wants to think about, yet a crisis is around the corner for every one of us in the construction industry. With the advent of the zero-injury philosophy, construction accidents have been dramatically reduced. But unfortunately, due to the high-hazard nature of the industry, a catastrophic event with a far-reaching impact on the corporate reputation and bottom line is always near. At Dick Corporation, we spend a significant amount of time planning our construction activities to ensure that safety is our first consideration. However, there is always the potential for hidden defects in construction materials or design deficiencies that could lead to a serious injury or incident. Because of this potential, every construction firm needs to be prepared for the worst.

The process starts with having a well-written crisis-management plan that addresses those serious situations that lie in wait for firms in the construction industry. Once a well-developed plan has been put together, an even greater challenge remains in communicating the plan to key personnel. Therefore, crisis-management planning is an ongoing activity for every construction firm.

—Andrew D. Peters
Vice President of Human Resources
Dick Corporation

When it comes to crisis management, writing the plan is the easy part. Training employees and keeping the plan updated and relevant for all of your various job conditions is, perhaps, more important. When a crisis occurs, the team needs to think about the unique aspects of the crisis at hand. Encourage employees to follow the plan, but to use their training and creative thinking to handle the specific situation in a proactive and professional manner.

—Joe A. Riedel, Jr.
Chairman and CEO
Beers Construction Company

As an insurance and surety broker with primary responsibility for risk-management consultation services for our construction clients, we know that a professionally prepared crisis-management plan is an essential element of our program to manage financial risk. Unfortunately, our experience in the construction industry has proven to us that it is not *if* but *when* a contractor will have need for crisis management and/or media-relations techniques. Simply put, failure to prepare is preparing to fail.

—Craig A. Merten, CPCU, ARM
President
The Linden Company

Sadly, there have been two major construction crises in our state recently. They both involved operations failures, worker injury, and significant problems for the contractors involved. There is no way to prepare for such an event, but it is absolutely critical to have your entire company ready to handle the crisis after it has occurred. These two crises have heightened our awareness and underscored the importance of our crisis-management plan.

Construction is a risky business loaded with potential crises every day, but firms that have crisis-management plans in place have an extreme advantage over those that do not. Controlling the situation and the general public's view of the company is at times reduced to a 30-second sound bite. It has to be done right! Furthermore, I am convinced our crisis-management plan instills a sense of awareness among our people in the field to avoid a crisis and to spot potential problems before they become real ones. This alertness to me is the greatest benefit of our plan—it helps us avoid crises before they happen.

—David J. Cullen
Vice President
J. P. Cullen & Sons, Inc.

As projects grow larger and schedules shorter, the possibility of a crisis becomes greater. With these elements becoming more common, companies need to be prepared for any crisis that might arise. Having a crisis-management plan and knowing what steps to follow could mean the difference between a crisis and a catastrophe.

—Bill Powell
Safety Director
Hoar Construction

Our tower-crane operator on a $75 million mall-renovation project had a massive heart attack and was unconscious upon discovery. Fortunately, I was on site, and we immediately began CPR and notified EMS. Within minutes, a large crowd consisting of employees and mall patrons began to gather. I worked closely with our team leader to determine how to remove the worker and handle crowd control at the same time. We also knew how critical it was to notify the injured worker's family, and contacting them became my priority as well as the company president's. On top of all of that, the print and broadcast media arrived just as the fire and rescue team was using the crane to lower the man-basket that contained our unconscious worker. We were interviewed by the media and were able to give a well-prepared statement that was sensitive to the family and workers involved. Unfortunately, the tower-crane operator died as a result of his heart attack. The president of our company was at the hospital with the operator's family and realized how important it was to lend support at this worst possible time in anyone's life. Had we not had Janine Reid's training earlier in the year and incorporated the plan she wrote for us, I am afraid we would not have handled the situation very well. Instead, Janine's words of advice were exactly what we needed to deal with a horrible situation.

In another situation, we had a fuel line break at an international airport, spilling 57,000 gallons of jet fuel onto the tarmac where the planes dock to load and unload passengers. Immediately, our crisis-management plan went into effect, and by late that evening the spill was cleaned up. The reports by the media focused on the great team effort that was made to complete the clean-up process. The media did not focus on the break or even mention our name! The plan allowed us to multi-task—handle the clean-up, owners, and media all at the same time and feel in control of the situation. The owners like the fact that we strive to avoid accidents, but are prepared to handle them in the best interest of all involved should they occur.

—Tracy L. Lawson, CHST
Director, Environmental Safety & Health
Fletcher Pacific Construction, Co., Ltd.

It has been our good fortune, in addition to hard work, to have avoided any major crises. We have, however, had several accidents that we have been able to control to the benefit of those involved as well as our clients and our own company. The ability to respond to these incidents is a by-product of our being knowledgeable and organized as the result of a crisis-management plan. Interestingly, we have had several incidents that did not involve injury per se, but were potentially highly disruptive, such as loss of power and fumes in a high-rise HVAC system, to name two. Because we were prepared and proactive, we were able to turn these incidents into powerful marketing events.

Our experience is that most building-owner clients are not well prepared for operations crises and that under these circumstances they become very vulnerable. On the other hand, we have been able to come in and take over a crisis situation, manage the emergency response teams as well as the media, and thereby avoid negative press for our clients and ourselves. One of these circumstances, eight years later, is still remembered and repeated by our clients as a high-value service not expected from us. While we have been lucky and need to stay ever-vigilant, we have seen real value to being prepared even in the absence of a real crisis.

—Stephen Jones
President
Snyder Langston

Preventing losses is the single most effective way to improve profit margins. It is as simple as that. In order to prevent losses from occurring, however, a company must be aware of its vulnerabilities and establish a protocol for practicing prevention on a daily basis.

It is an unfortunate fact of life that accidents can and do happen—even with the best safety programs. This fact drives home the importance of having a crisis-management plan and implementing a training program on what to do in the event of an accident. This process not only creates an increased level of awareness for prevention, but also guides the field and corporate office through all of the necessary steps that must be taken to ensure a successful outcome to an unexpected situation. It is my hope and dream that a company's crisis-management plan will never be used, but one feels a certain comfort knowing it is available—just in case.

—Sherwood Kelly
Senior Vice-President
Director of Safety Management
Willis Construction Practice

A contractor relies heavily upon reputation and the public's perception of the company's ability to achieve the community's business goals. This reputation is built over the course of many years and many projects. A crisis has the potential to cause great harm to a company's reputation, particularly if the crisis is mishandled.

A well-developed and properly implemented crisis-management plan is an essential step to protect this hard-earned reputation. As an additional benefit, the training, which is required to implement a plan, presents the potential consequences of a crisis and thereby assists in reinforcing the importance of safe work practices.

—David K. White
Vice President/Denver Operations
Swinerton & Walberg

It seems to me that, in addition to protection of one's bottom line, a major reason to have in place and use a crisis-management plan is the fulfillment of one's professional obligations. A professional constructor has a duty to the safety and well-being of his or her employees as well as to the general public. If and when "the sky falls," part of that ethical obligation is to prevent further injury or risk effectively and honestly and to inform the public (frequently through the press) of what is known at the time. I believe this is one of those occasions when one's professional responsibility aligns with what is right for the bottom line. As this book suggests, a truthful and immediate response with the facts known at the time—a response that seeks neither to establish nor to avoid blame— is always a better approach than the temptation to say, "No comment." Forthrightness is not only the best pragmatic approach, but also the appropriate professional, ethical response.

—Al Hauck, PhD, CPC, AIC
Associate Professor and Undergraduate Program Coordinator
Construction Management Program
Colorado State University

There are some unique instances when a short-term financial gain can be a by-product of a crisis, but a company's reputation in the marketplace can suffer if the media portrays the company in a negative light. If a company's reputation is stained as a result of this attention, the recovery process can be both costly and time-consuming.

Relationships are vital in the construction industry, and in a time of crisis a company under siege must do whatever is necessary to protect its reputation and position in the marketplace. This protection can be accomplished only through quick and consistent communication. A company's reputation and relationships are fragile; without some type of action plan, its hard-earned reputation can vanish in a very brief time.

—Duane Pozza
Principle-in-Charge
Bartlett Cocke, L.P.

People usually do not give any thought to what they would say or do in a crisis. If and when one does occur, they find that they are inadequately prepared and more often than not blurt out misinformation in answer to questions asked under stressful conditions. After our employees received training in this area, they commented on how surprised they were to learn how "rattled" their thinking could become in chaotic and tense situations. We feel that both training and a crisis-management plan are necessary to give employees direction and confidence.

—Delaine Nelson
Director
MRM Group, Inc.

King and Neel is a locally owned insurance agency and surety bonding company in Honolulu. We have embraced the discipline of crisis-management planning with our construction clients since 1991, and the decision has clearly paid off. Shortly after the managers of one of our largest general contractors attended Janine Reid's training session, a fire broke out in the early morning hours at a shopping center that was undergoing a $75 million renovation. The contractor's crisis-management plan was implemented, and all of the involved staff members knew what to do—and did it without hesitation.

There was no publicity except for one of the retail stores advertising a fire sale! It is doubtful that the outcome of these events would have been as positive without the no-nonsense guidance that Janine's plan and training provided.

—Patrick J. Conroy
Vice President, Risk Control Services
King and Neel, Inc.

There are two events that stand out in my mind about crisis management. The first occurred while watching the 6:00 P.M. news. I saw one of my client's senior officers literally running from the press over a breaking news item about my client's operations. Apparently, he tried to sneak out the back door, and the TV press crew caught him. He did not know what to say, and all I can tell you is that he looked really bad on local TV. In fact, when the investigation was complete, our client was totally cleared; however, in my mind I can still see that man running from the camera because he did not know how to handle the situation.

The other event was the time when I attended the IRMI [International Risk Management Institute] Convention, where Janine Reid was one of the key speakers. I remember thinking, "Why not sleep late, because who cares about crisis management?" But, for whatever reason, I got up and Janine's seminar changed my opinion about this topic. I really believe that it is not a matter of *if* a corporation or entity will have a problem with the press, but *when*. I used to think that it was a fact that the press was biased and always worked towards their own agenda to get a sensational story. I still believe that I am still partially correct, but I now believe that you have to take a proactive position to get your side of the story told. If you are not trained in handling the press, you are doomed and could appear guilty, even when you are not.

—W. Meade Collinsworth, PCU, ARM, AIM, AAI
Collinsworth, Alter, Nielson, Fowler and Dowling, Inc.

Fru-con believes in the value of being prepared. Due to Janine's training, we know that when an unplanned or unpleasant event occurs, our supervision and leadership are confident in their own ability to handle a crisis in an organized and efficient manner. In our business, our private customers are essential to our future; and when they see us handling adversity in a professional and effective manner, they feel more confidence in our abilities. Conversely, if our customers saw us stonewalling the media or sensed that we were not prepared to handle a crisis, they would tend to not include us in their construction plans.

The keys to responding to a crisis are being organized and communicating well. Being organized allows us the opportunity to anticipate and be proactive in our response—steps that to me are the hallmark in minimizing any loss.

—Elbe Watkins
Corporate Safety Director
Fru-Con Construction Corporation

Preparing for the unexpected is crucial in the construction industry because it happens every day. The difference between routine unexpected situations and a major crisis is the level of damage, injury, commotion, disruption, hostility, or media attention the event provokes. Good managers deal with the routine situations every day because they fit into the regular planning process. However, good managers often fail horribly in a major crisis while events swirl out of control. As with all of our work, having a solid plan is critical. Most of us would not send a crew with equipment and materials out at the start of a shift without a plan. Why should we do less for crisis situations?

Development of a project-specific crisis-management plan should be a part of every major project. The key to these plans is a clear assignment of responsibility, notification requirements, and a checklist of actions to take. Once prepared, you must train and retrain everyone. Classroom sessions as well as video and role-playing exercises are all good activities. Crisis planning will not eliminate the crises, but it will help your managers deal with these situations and reduce their impacts on your people and your company.

—Peter Miller
President
Kiewit Network Services

APPENDIX

SURVEY RESULTS

This survey was intended to cover a broad cross section of the construction industry, both in terms of dollar value and work performed.

Profile of Companies

	1996—149 Responses	1988—125 Responses
	1996 % Respondents	1988 % Respondents
General contractors	48%	53%
Heavy/Highway	19%	21%
Subcontractors	17%	22%
Engineering/Construction	7%	4%
Manufacturer	1%	0
Owner	3%	0
Aggregate producers	1%	0
Ready-mix Suppliers	4%	0

General Contractor

	1996—71 Responses	1988—67 Responses
Annual company sales for 1995	1996 % Respondents	1988 % Respondents
Under $1 million	1%	0
$1–5 million	6%	0
$5–15 million	21%	6%
$15–30 million	4%	21%
$30–50 million	10%	24%
$50–100 million	20%	26%
$100–200 million	17%	9%
Over $200 million	21%	14%

Heavy/Highway

	1996—29 Responses	1988—26 Responses
Annual company sales for 1995	1996 % Respondents	1988 % Respondents
Under $1 million	0	0
$1–5 million	10%	0
$5–15 million	28%	4%
$15–30 million	14%	4%
$30–50 million	10%	23%
$50–100 million	21%	46%
$100–200 million	17%	15%
Over $200 million	0	8%

Subcontractors

	1996—26 Responses	1988—27 Responses
Annual company sales for 1995	1996 % Respondents	1988 % Respondents
Under $1 million	4%	0
$1–5 million	8%	0
$5–15 million	19%	4%
$15–30 million	19%	18%
$30–50 million	22%	30%
$50–100 million	12%	37%
$100–200 million	8%	7%
Over $200 million	8%	4%

Manufacturer

	1996—1 Response	1988—0 Responses
Annual company sales for 1995	1996 % Respondents	
Over $200 million	100%	

Engineering/Construction

	1996—10 Responses	1988—5 Responses
Annual company sales for 1995	1996 % Respondents	1988 (reflects '87 volume) % Respondents
Under $1 million	0	0
$1–5 million	0	0
$5–15 million	10%	0
$15–30 million	10%	0
$30–50 million	0	0
$50–100 million	10%	20%
$100–200 million	40%	20%
Over $200 million	30%	60%

Owner

	1996—4 Responses	1988—0 Responses
Annual company sales for 1995	1996 % Respondents	
Over $200 million	100%	

Aggregate Producer

	1996—2 Responses	1988—0 Responses
Annual company sales for 1995	1996 % Respondents	
$50–100 million	50%	
Over $200 million	50%	

Ready-Mix Supplier

	1996—6 Responses	1988—0 Responses
Annual company sales for 1995	1996 % Respondents	
$1–5 million	34%	
$5–15 million	50%	
$100–200 million	16%	

The participants were asked to indicate crisis situations to which they were exposed within the last three years. They were presented with a list of 27 possible situations to choose from. The most frequently mentioned crises follow.

Types of Crises—1996

Rank		Number of Responses
1	On-the-job accident requiring hospitalization	94
2	Damage to utility lines	73
3	Contractual disputes with client resulting in litigation	64
4	Equipment failure	61
5	On-the-job fatality	45
6	Highway accident	44
7	Theft/embezzlement	42
8	Noise/dust pollution	35
9	Sexual harassment	33
10	Labor strike/work stoppage	32
11	Complaints to the media about your company	29
12	Fire/explosion	28
13	Community/environmental protests	27
14	Environmental violations	25
15	Structural/Subsidence collapse	24
16	Merger/acquisition	18
17	Workforce violence	17
18	Serious cash flow problems	16
19	Sabotage	15
20	Loss of key supplier	14
21	Bomb threats	12
22	Loss of bonding capability	9
23	Death of owner or key employee	9
24	Scandal involving top management	9
25	Sexual discrimination	8
26	Long-term structural problems	7
27	Bid-rigging accusations	4

Other crises mentioned were death threat to owner, drug involvement, flood, non-payment on a major project, corporate aircraft seizure, lack of a qualified workforce; and race discrimination.

Types of Crises—1988

Rank		Number of Responses
1	On-the-job accident requiring hospitalization	101
2	Contractual disputes with client resulting in litigation	81
3	Damage to utility lines	64
4	On-the-job fatality	58
5	Theft/embezzlement	39
6	Labor strike or work stoppage	33
7	Serious cash flow problems	33
8	Rapid growth	32
9	Lack of bonding capability	23
10	Sudden market shift	23
11	Environmental violations	22
12	Complaints to the media about your company	17
13	Merger/acquisition	16
14	Loss of key supplier	15
15	Sex discrimination	14
16	Sabotage	13
17	Structural collapse	8
18	Long-term structural problems	5
19	Death of owner or key employee	5
20	Payoff/kickback accusations	5
21	Owner won't retire	3
22	Bomb threats	3
23	Loss of critical computer data	2
24	Scandal involving top management	2

Other crises mentioned were minority and age discrimination, bankruptcy of client, cancellation of insurance, allegations from the Nuclear Regulatory Commission, availability and cost of liability insurance.

REFERENCES

Acid injures workers at refinery where 6 died. (1999, April 15). *Seattle Times,* p. B1.

Barton, L. (1993). *Crisis in organizations.* Cincinnati, OH: South-Western.

Bureau of Labor Statistics. (1999, August 12). *1998 national census of fatal occupational injuries* [On-line]. Available: http://www.bls.gov/special. requests/ocwc/oshwc/cfoi

Caponigro, J. R. (1998). *The crisis counselor.* Southfield, MI: Barker Business Books.

Construction worker buried alive in pit. (1998, July 12). *The New York Times,* p. 25.

Colorado's constructors build a living legacy. (1999, August 16). *ENR,* p. 100.

Fatal cave-in at work site raises questions about safety; family says worker had misgivings before tragedy in Antonia. (1999, February 4). *St. Louis Post-Dispatch,* p.1.

Fink, S. (1986). *Crisis management: Planning for the inevitable.* New York: American Management Association.

Grossly incompetent design central to walkway disaster. (1999, March). *Safety & Health Practitioner,* p. 8.

Hinze, J., & Applegate, L. L. (1991, September). Cost of construction injuries. *Journal of Construction Engineering and Management, 117,* 537.

Lerbinger, O. (1997). *The crisis manager.* Mahwah, NJ: Lawrence Erlbaum.

Lukaszewski, J. E. (1994, January/February). 2010: An age of specialty. *IABC Communication World, 11,* 37.

Lukaszewski, J. E. (1999). Regaining credibility following a damaging situation. *Executive Action.* Used with permission of the author.

Mills, S. (1991, March/April). Salons and beyond: Changing the world one evening at a time. *Utne Reader, 44,* 68–77.

Mindszenthy, B., Watson, T. A. G., & Koch, W. (1988). *No surprises.* Toronto: Bedford House.

Mitchell, J., & Everly, G. (1996). *Critical incident stress debriefing: An operations manual for the prevention of traumatic stress among emergency services and disaster workers.* Ellicott City, MD: Chevron.

Oakland cave-in kills worker: Fellow laborer decapitates body in rescue attempt. (1999, April 30). *The Detroit News*, p. C4.

Peters, T. (1994). *The pursuit of WOW! Every person's guide to topsy-turvy times.* New York: Vintage Books.

Pew Research Center for the People & the Press. (1999, February). *1999 Journalists Survey: Overview* [On-line]. Available: http://www.people-press.org/press99rpt.htm

Scanlon, J. (1999, March 8). Scanlon: Monica's new man. *PR Week, 2,* 18.

Sheet Metal & Air Conditioning Contractors' Association. (1998, February). Hidden costs of accidents on the job site can bankrupt a company. *Industrial Insights, 2,* 5.

Third fatality strikes Las Vegas casino. (1999, March 15). *ENR*, p. 9.

Utility crew punctures gas line, explosion kills 4. (1998, December 21). *ENR*, p. 7.

Walsh, B. (1993, March 29). When things go bad. *Forbes*, ASAP Supplement, 13–14. (Reprinted by permission of FORBES ASAP Magazine © Forbes Inc., 1993).

Worker is killed in collapse linked to ill-rigged scaffold (1996, August 6). *The New York Times*, p. B1.

INDEX